GILBERT SIMONDON

On the Mode of
Existence of Technical Objects

Translated by Cécile Malaspina and John Rogove

A Univocal Book

University of Minnesota Press

Minneapolis

London

Univocal Publishing was founded by Jason Wagner and Drew Burk as an independent publishing house specializing in artisanal editions and translations of texts spanning the areas of cultural theory, media archeology, continental philosophy, aesthetics, anthropology, and more. In May 2017, Univocal ceased operations as an independent publishing house and became a series with its publishing partner the University of Minnesota Press.

Du mode d'existence des objets techniques
by Gilbert Simondon
Copyright Aubier (department of Flammarion) Paris, 1958 and 2012

Originally published by Univocal Publishing, 2017
First University of Minnesota Press Edition 2017.

Published by the University of Minnesota Press
111 Third Avenue South, Suite 290
Minneapolis, MN 55401-2520
http://www.upress.umn.edu

ISBN 978-1-5179-0487-6

A Cataloging-in-Publication record for this book
is available from the Library of Congress.

Printed in the United States of America on acid-free paper

The University of Minnesota is an equal-opportunity educator and employer.

27 26 25 24 23 22 21   10 9 8 7 6 5 4 3 2

TABLE OF CONTENTS

ON THE MODE OF EXISTENCE OF TECHNICAL OBJECTS

PART I
GENESIS AND EVOLUTION OF TECHNICAL OBJECTS

CHAPTER ONE
Genesis of the technical object: the process of concretization

PART III
The Essence of Technicity

CHAPTER ONE
The genesis of technicity

CHAPTER TWO
Relations between technical thought and other species of thought

CHAPTER THREE
Technical and philosophical thought

NOTE

The intention of this note is not to provide a presentation of *On The Mode of Existence of Technical Objects*, a work that has for quite some time already become a classic philosophical reference and is Gilbert Simondon's most well known work. Rather what is provided here is a number of informational comments on Gilbert Simondon's career as well as his other, more recent, and lesser-known publications. The best presentation of *Mode of Existence* is, of course, the *Prospectus* created by the author for the first edition published in 1958 in Paris and which appears in its entirety after this note. The reader will also find, at the end of this present note, a "summary" of the book that was written later by the author himself.

Gilbert Simondon (1924-1989) was a student at the École Normale Supérieure (rue d'Ulm in Paris) from 1944-1948. Attaining the status of *agrégé de philosophie*, he began his career teaching philosophy at the *lycée de Tours* (teaching philosophy to high school students as is still a common practice in the French educational system) and later on taught at the University of Poitiers. He defended his doctoral thesis [*thèse de doctorat*] in 1958 and was then named professor at the Sorbonne in 1963. Only his two principal works are published while he is a living philosopher, *On The Mode of Existence of Technical Objects* and *L'Individuation à la lumière des notions de forme et d'information* [*Individuation in light of notions of form and information*].

Since then, his other writings, lectures, and conference papers, written between 1950 and 1985, have been assembled into a dozen more published works. *Deux leçons sur l'animal et l'homme, L'invention dans les techniques, Cours sur la perception, Imagination et invention, Communication et information, Sur la technique, Sur la psychologie, Sur la philosophie*, and finally *la Résolution des problèmes*.

The career of Gilbert Simondon was that of a typical academic and researcher. It became the perpetual deployment of the ideas and concepts contained in the first two published works, as much within the lectures and notes of his yearly courses he taught at the university, as in the research he actively undertook in the laboratory he directed at the Sorbonne (and later on when the Sorbonne was divided into other branches, at the Université Paris V). Through the study of invention, perception, and imagination, this research develops the preoccupations that led him, as a philosopher and scientist, to grasp the problem of information in its relation with individuation, at the beginning of the 1950's.

How does something appear, how is something capable of becoming individuated, how does a given structure, a given form emerge? How, on the other hand, can it be grasped by thought? This problem, that is in no way new but by way of Simondon's approach to it is renewed through a critique of the concept of hylomorphism (matter and form) that masks the principal point of individuation, is at the heart of everything: of physics, of technics, of the problems that technics poses to human beings, of the living, and in the end the individuation of man himself, in as much for thinking the psycho-physiological relation (and to grasp in another way the problem of relation of the soul and the body) as for thinking the social and the collective and, as a result, rethinking the human sciences. As the author himself notes in preparatory notes, this is an ontological and epistemological problem, which also involves a value judgment, that is an "axiontological" problem. It is a "reflexive" problem, i.e., a philosophical problem. Gilbert Simondon's intention is to make people understand that a true individual is not simply something that is individuated, enclosed within his separate being, but is rather a theatre for new individuations. The collective, if it is not conflated with the purely social, is the site of this liberation and of a veritable progress. As such, "ethics is the direction [sens] of individuation" and "societies become a world."[1]

These works were conducted throughout the 1950s with a full awareness of the other philosophies in the midst of development, and with a full awareness of the need for philosophy to be connected to life in all its varied dimensions (what Gilbert Simondon calls "the incompletion of non-reflexive life" [*l'inachève-ment de la vie non-réflexive*]). But no philosophical school seemed to provide the means for what is required to achieve this: "A reflexive attitude must begin by avoiding to postulate a specific determined end or affiliation [*appartenance*] at the very moment when it begins to exist and strives to define itself. A philosophy that would accept being defined by way of some qualifier such as "Christian", "Marxist", or "Phenomenological" would find the negation of its philosophical nature within this initial determination" ("Note sur l'attitude reflexive", in *Sur la philosophie*). Existentialism is itself reckless and lacking in prudence: "It is perfectly

---

1. From the conclusion of *L'Individuation à la lumière des notions de forme et d'information.*

true that a vital engagement is an inexhaustible source of rigor [*de sérieux*], but it is uncertain whether this quality of authenticity can be directly transposed into an explicit thought according to an intellectual systematic that is already classified and known. The explicit translation of an engaged implicit thought runs the risk of being abstract in spite of the force of the concrete position it wants to express: nothing can guarantee the authenticity of the transposition." If philosophy must bear the weight of the concern for the concrete, it's by way of a long detour of the most radical analysis of things, by way of the study of ontogenesis and the conditions of an individuation (placing into relation form, information, and potential). This analysis alone has the power of unmasking the psychosocial myths that serve as political theories.

And the same can be said for technics as well, the study must be a genetic one, and allow itself to leave behind the false categories (of genera and species) to which we reduce technical objects when we think of them starting from the way they are used. Individuation must be studied within the technical domain, alongside physical individuation as well as psychical and biological [*vital*] individuation: for the very reason that the technical object "is that from which there is a genesis". Technics is even a domain in which individuation plays itself out in a remarkable fashion, in the case of technical invention where a mental operation and a technical operation coincide. Part III of this present book (The Genesis of Technicity) locates technics within a more global perspective of a genetic theory of culture. The dimension that Gilbert Simondon grants to the study of technics, in making it the unique object of his complementary thesis is already justified in the introduction of *On The Mode Of Existence of Technical Objects* by way of a principal issue, linked to the problem technics poses to culture: the technical object, that impoverished relative of culture, is not granted the same dignity that is, in contrast, conferred to the aesthetic object. As we see emphasized in the "Note complémentaire sur les conséquences de la notion d'individuation" (from ILFI), technics is a fundamental issue that is in no way whatsoever foreign to the general preoccupations concerning psychic and collective individuation. For the very reason that technics is the way in which man is in relation with the world, and not simply with a community that is closed in on itself. Hence the importance of invention, that loosens or frees the community through opening it up to a true and free society, and the importance of the welcoming of technics into and by culture: technics is not, for the philosopher, simply an object among others, technics constitutes a field of reflection that philosophy must plunge into and become invested [*investir*]. A lucid gaze focused on technical objects must first and foremost delineate their diverse modes of existence (element, individual, ensemble) if one wants culture, in its relation to technics, to be the bearer of freedom [*liberté*] and not alienation. Certain prolonged analyses that began in *On The Mode of Existence of Technical Objects*, developed in chapters that were not integrated into the original version of the thesis, have now become available in the published volume *Sur la philosophie* (Paris: PUF, 2016): see for instance "The technical object as paradigm of universal intelligibility,"[2] "The order of technical objects as axiological universal paradigm in interhuman relation (Introduction to a transductive philosophy)."[3]

The later works of Gilbert Simondon concerning technics continue the construction site opened up by the early work that takes place in *On The Mode of Existence of Technical Objects*: there are numerous texts concerning the social and cultural modes of existence of technical objects (for example *Psychosociologie de la technicité*) as well as the problem of an adequate relation between man and technical objects. Other texts are more concerned with the conditions of true progress, and still others are more focused on invention, in its objective aspect (and the conditions favoring its emergence) and in its mental aspect. More generally, the works on perception, imagination, and communication are driven by the same perspective that Gilbert Simondon held onto from the very first of his writings: to understand these processes by way of grasping what individuates, to construct a reflexive path (even, and above all within the area of psychology) that never gives up on "optative" (the hoped for), i.e., that by which every problem that is posed to man can be thought and reflected upon according to its true dimensions, and, as such, should then be resolved.

Nathalie Simondon

Summary of *On The Mode Of Existence of Technical Objects*
by Gilbert Simondon

This present work is concerned with the essence of technical objects and their relation to man. While the aesthetic object has been considered suitable material for philosophical reflection, the technical object, treated as an instrument, has only

---

2   "L'objet technique comme paradigme d'intelligibilité universelle".

3   "L'ordre des objets techniques comme paradigme d'universalité axiologique dans la relation interhumaine (Introduction à une philosophie transductive)".

ever been directly studied across the multiple modalities of its relation to man as an economic reality, as an instrument of work, or, indeed, of consumption.

The nonessential character of knowledge of the technical object with respect to its different relations to man has contributed to masking a task incumbent upon philosophical thought: to rediscover, through a deepening of the relation which exists between nature, man, and technical reality, the burden of alienated human reality which is enclosed within the technical object. The technical object, taking the place of the slave and being treated as such across relations of property and custom, has only partially liberated man: the technical object possesses a power of alienation because it is itself in a state of alienation, one more essential than economic or social alienation.

The importance of technical objects in contemporary cultures requires philosophical thought to make the effort of reducing technological alienation by introducing into culture a representation and scale of values adequate to the essence of technical objects.

The discovery of this essence must be carried out through a study of the genesis of technical objects, achieving itself [*s'accomplissant*] through a process of concretization which is different from successive empirical corrections and from deduction from prior theoretical principles: there is a specific genesis of the technical object.

A historical study allows for the discovery of the regulative function of culture in the relation between man and the technical object, especially across the normative foundation of the successive manifestations of encyclopedic spirit, from the technicism of the Sophists up to cybernetic theory, passing through the open and the autonomous awareness of the work of Diderot and d'Alembert.

Finally, a study of the contemporary modalities of the relation between man and the technical object shows that the notion of information is the most suitable for accomplishing the integration of culture with a representative and axiological content adequate to technical reality envisaged in its essence, man becoming, after, invention, the active center and actor who alone can bring into existence a coherent technological world.

PROSPECTUS
Presentation written in 1958 by Gilbert Simondon

The book titled *On the Mode of Existence of Technical Objects* aims at introducing a knowledge into culture that is adequate to technical objects considered on three levels: elements, individuals, ensembles. A gap manifests itself in our civilization between the attitudes provoked in man by the technical object and the true nature of these objects; from this inadequate and confused rapport a set of mythological valuations and devaluations arises in the consumer, the manufacturer, and the worker; in order to replace this inadequate rapport with a veritable relation, one has to become aware of the mode of existence of technical objects.

This becoming aware takes place in three stages.

The first seeks to grasp the genesis of technical objects: the technical object mustn't be seen as an artificial being; the sense of its evolution is a concretization; a primitive technical object is an abstract system of isolated partial ways of functioning, without common ground of existence, without reciprocal causality, without internal resonance; a perfected technical object is an individualized technical object in which each structure is pluri-functional, overdetermined; in it each structure exists not only as organ, but as body, as milieu, and as ground for other structures; in this system of compatibility whose systematicity [*systématique*] takes form just as an axiomatic saturates, each element fulfills not only a function in the whole [*ensemble*] but a function of the whole. There is something like a redundancy of information in the technical object having become concrete.

This notion of information allows the general evolution of technical objects to be interpreted via the succession of elements, of individuals and of ensembles, according to the law of conservation of technicity. The veritable progress of technical objects takes place through a schema of relaxation and not of continuity: there is a preservation throughout the successive cycles of evolution of technicity as information.

xvi

The second phase envisions the rapport between man and the technical object, on the one hand at the level of the individual, and on the other hand, at the level of ensembles. The individual's mode of access to the technical object is *minor* or *major*; the minor mode is the mode appropriate for the knowledge of the tool or the instrument; it is primitive, but adequate to this level of the existence of technicity in the form of tools or instruments; it turns man into a bearer of tools, according to a concrete apprenticeship, a sort of instinctive symbiosis of man and the technical object employed in a determinate milieu, according to intuition and implicit, almost innate knowledge. The major mode presupposes the becoming aware of the ways of functioning: it is polytechnic. Diderot and Alembert's *Encyclopedia* illustrates the passage from the minor to the major mode.

At the level of ensembles, the awareness that the group gains from its rapport with technical objects is translated by way of diverse modes of the notion of progress, which are the various value judgments made by the group regarding the power harbored by technical objects in order to facilitate the evolution of the group: the optimistic progress of the 18th century corresponds to the awareness of the improvement of elements; the pessimistic and dramatic progress of the 19th century corresponds to the replacement of the individual human tool bearer by the machine individual, as well as corresponding to the anxiety resulting from the frustration of this progress. Finally, what remains to be elaborated is a new notion of progress corresponding to the discovery of technics at the level of the ensembles of our epoch, by virtue of a deepening of the theory of information and communication: the true nature of man is not to be a tool bearer — and thus a competitor of the machine, but man's nature is that of the inventor of technical and living objects capable of resolving problems of compatibility between machines within an ensemble; he coordinates and organizes their mutual relation at the level of machines, between machines; more than simply governing them, he renders them compatible, he is the agent and translator of information from machine to machine, intervening within the margin of indeterminacy harbored by the open machine's way of functioning, which is capable of receiving information. Man constructs the signification of the exchanges of information between machines. The inadequate rapport of man and the technical object must therefore be grasped as a coupling between the living and the non-living. Pure automatism, excluding man and aping the living, is a myth that does not correspond to the highest level of possible technicity: there is no machine of all machines.

Finally, the third phase of becoming aware, places the technical object back *into the ensemble of the real*, seeking to know the technical object according to its essence, according to a genesis of technicity. The basic hypothesis of the employed philosophical doctrine consists in supposing the existence of a primitive mode of man's relation with the world, which is the magical mode: from an internal rupture of this relation arise two simultaneous and opposite phases, the technical phase and the religious phase; technicity is the mobilization of the figural functions, the extraction of the key-points of man's relation with the world; religiosity on the contrary refers to the respect for the ground functions: it is the attachment to the totality in its ground. *This relation of phase shifts of man to the world obtains an imperfect mediation via aesthetic activity*: aesthetic thought preserves the nostalgia of man's primitive relation to the world; it is a neutrality between opposing phases; but its concrete character as constructor of objects limits its power of mediation, for the aesthetic object loses its neutrality, and consequently its power of mediation by seeking to become either functional or sacred. It is only at the level of both the most primitive and the most elaborate of all thoughts, philosophical thought, that a truly *neutral* and *balanced* because *complete,* mediation between opposing phases can intervene. It is thus *philosophical thought* alone that can assume the knowledge, valorization and completion of the phase of technicity within the entirety [*ensemble*] of man's modes of being in the world, by way of a meditation regarding the rapport between science and technics, theology and mysticism.

ON THE MODE OF EXISTENCE OF TECHNICAL OBJECTS

ACKNOWLEDGMENTS

I want to thank my former professors, André Bernard, Jean Lacroix, Georges Gusdorf, and Jean-T. Desanti.

I also want to express my sincere gratitude to my dear old colleagues, André Doazan, and Mikel Dufrenne, who were present in Paris to support me during the defense of my dissertation.

In particular, I want to thank M. Dufrenne for his repeated words of encouragement, for his helpful suggestions, and for his active sympathy during the editing and revisions of this study.

M. Canguilhem, obligingly allowed for me to obtain a number of documents from the library of the L'Institut de l'histoire des sciences and lent me a number of rare works written in German from his own personal library. Moreover, it was thanks to all of M. Canguilhem's helpful remarks that I was finally able to discover the definitive form of this work; the third part of the present work owes a great deal to his suggestions. I want to publically express my sincere recognition for his firm generosity.

G.S.

Introduction

This study is motivated by our desire to raise awareness of the meaning of technical objects. Culture has constituted itself as a defense system against technics[1]; yet this defense presents itself as a defense of man, and presumes that technical objects do not contain a human reality within them. We would like to show that culture ignores a human reality within technical reality and that, in order to fully play its role, culture must incorporate technical beings in the form of knowledge and in the form of a sense of values. Awareness of the modes of existence of technical objects must be brought about through philosophical thought, which must fulfill a duty through this work analogous to the one it fulfilled for the abolition of slavery and the affirmation of the value of the human person.

'The opposition drawn between culture and technics, between man and machine, is false and has no foundation; it is merely a sign of ignorance or resentment. Behind a facile humanism, it masks a reality rich in human efforts and natural forces, and which constitutes a world of technical objects as mediators between man and nature.

Culture behaves toward the technical object as man toward a stranger, when he

---

1.   The broad key term, "la technique," and its plural, "les techniques," are translated uniformly throughout by the more specialized "technics," meaning the theory or study of industry and of the mechanical arts; while this term, as a collective plural used in the singular along the same lines as "physics," is usually a near synonym to "technology" and is differentiated in English from "technique" insofar as the latter refers to the almost ineffably practical and particular application of technics to a given concrete task, in French the singular "la technique" and the plural "les techniques" cover together the meanings covered both by "technique" or "techniques" in English and by "technics," and so the word "technics" as it appears in this text accordingly covers both. Moreover, Simondon is careful to distinguish "technics" from "technology," which remains programmatic in Simondon's text and the elaboration of which, as a philosophical logos or meta-theory of technics, this text may be construed as an outline. [TN]

10 allows himself to be carried away by primitive xenophobia. Misoneism directed against machines is not so much a hatred of novelty as it is a rejection of a strange or foreign reality. However, this strange or foreign being is still human, and a complete culture is one which enables us to discover the foreign or strange as human. Furthermore, the machine is the stranger; it is the stranger inside which something human is locked up, misunderstood, materialized, enslaved, and yet which nevertheless remains human all the same. The most powerful cause of alienation in the contemporary world resides in this misunderstanding of the machine, which is not an alienation caused by the machine, but by the non-knowledge of its nature and its essence, by way of its absence from the world of significations, and its omission from the table of values and concepts that make up culture.

Culture is unbalanced because it recognizes certain objects, like the aesthetic object, granting them citizenship in the world of significations, while it banishes other objects (in particular technical objects) into a structureless world of things that have no signification but only a use, a utility function. Confronted by such a defensive rejection, pronounced by a partial and biased culture, men who have knowledge of technical objects and who appreciate their signification seek to justify their judgment by granting the technical object the only status currently valued besides that of the aesthetic object, namely that of the sacred object. This, then, gives rise to an intemperate technicism which is nothing other than idolatry of the machine and which, through this idolatry, by means of identification, leads to a technocratic aspiration to unconditional power. The desire for power consecrates the machine as a means of supremacy, it makes of it a modern philter. The man who wants to dominate his peers calls the android machine into being. He thus abdicates before it and delegates his humanity to it. He seeks to construct a thinking machine, dreams of being able to build a volition machine, a living machine, in order to retreat behind it without anxiety, freed of all danger, exempt from all feelings of weakness, and triumphant through the mediation of what he invented. In this case, however, the machine, after having become, according to the imagi-
11 nation, the robot, this duplicate of man devoid of interiority, quite evidently and inevitably represents a purely mythical and imaginary being.

We would like to show, precisely, that the robot does not exist, that it is not a machine, no more so than a statue is a living being, but that it is merely a product of the imagination and of fictitious fabrication, of the art of illusion. The notion of the machine as it currently exists in culture, however, incorporates to a great extent this mythical representation of the robot. An educated man would never dare to speak of objects or figures painted on a canvas as genuine realities, having interiority, good or ill will. However, this same man speaks of machines as threatening man, as if he attributed a soul and a separate, autonomous existence to them, conferring on them the use of sentiment and intention toward man.

Culture thus has *two contradictory attitudes* toward technical objects: on the one hand, it treats them as pure *assemblages of matter,* devoid of true signification, and merely presenting a utility. On the other hand, it supposes that these objects are also robots and that they are animated by hostile *intentions* toward man, or that they represent a permanent danger of aggression and insurrection against him. And judging it better to cling to the first characteristic, it seeks to prevent the manifestation of the second and speaks of placing machines in the service of man, in the belief that the reduction to slavery is a sure way to prevent any rebellion.

This inherent contradiction within culture in fact comes from the ambiguity of the ideas related to automatism, and in them we discover the hidden logical flaw. Worshipers of the machine commonly present the degree of perfection of a machine as proportional to the degree of automatism. Going beyond what experience shows, they suppose that by increasing and perfecting automatism one would manage to combine and interconnect all machines among themselves, in such a way as to constitute the machine of all machines. 12

Automatism, however, is a rather low degree of technical perfection. In order to make a machine automatic, one must sacrifice a number of possibilities of operation as well as numerous possible usages. Automatism, and its utilization in the form of industrial organization, which one calls *automation,* possesses an economic or social signification more than a technical one. The true progressive perfecting of machines, whereby we could say a machine's degree of technicity is raised, corresponds not to an increase of automatism, but on the contrary to the fact that the operation of a machine harbors a certain margin of indeterminacy. It is this margin that allows the machine to be sensitive to outside information. Much more than any increase in automatism, it is this sensitivity to information on the part of machines that makes a technical ensemble possible. A purely automatic machine completely closed in on itself in a predetermined way of operating would only be capable of yielding perfunctory results. The machine endowed with a high degree of technicity is an open machine, and all open machines taken together [*l'ensemble des machines ouvertes*] presuppose man as their permanent organizer, as the living interpreter of all machines among themselves. Far from being the supervisor of a group of slaves, man is the permanent organizer of a society of technical objects that need him in the same way musicians in an orchestra need the conductor. The conductor can only direct the musicians because he plays the piece the same way they do, as intensely as they all do; he tempers or hurries them, but is also tempered or hurried by them; in fact, it is through the conductor that the members of the orchestra temper or hurry one another, he is the moving and current form of the group as it exists for each of them; he is the mutual interpreter of all of them in relation to one another. Man thus has the function of being the permanent coordinator and inventor of the machines that surround him. He is *among* the machines that operate with him.

Man's presence to machines is a perpetuated invention. What resides in the machines is human reality, human gesture fixed and crystallized into working structures. These structures need support during the course of their operation, and the greatest perfection coincides with the greatest openness, with the greatest freedom of operation. Modern calculating machines are not pure automata; they are technical beings that, beyond their automatisms of addition (or of decision according to the operation of elementary switches), possess a great range of possibilities for the switching circuits, which allow for the coding of the machine's operation by reducing its margin of indeterminacy. This primitive margin of indeterminacy is what allows the same machine to extract cube roots or to translate a simple text, composed of a small number of words and expressions, from one language into another.

It is also through the intermediary of this margin of indeterminacy and not through automatisms that machines can be grouped into coherent ensembles and exchange information with one another via the intermediary of the coordinator that is the human interpreter. Even when two machines exchange information directly (as between a master oscillator and another pulse synchronized oscillator), man intervenes as a being who regulates the margin of indeterminacy in order to adapt it to the best possible exchange of information.

Now, one might wonder who can achieve within himself an awareness of technical reality and introduce it to culture. This awareness can hardly be achieved by someone who is attached to a single machine through work and the fixity of daily gestures; the use relation is not conducive to the raising of awareness, because its habitual repetition erases the awareness of structures and operations with the stereotypy of adapted gestures. Running a company that uses machines, or owning one, is no more useful for the attainment of this awareness than is labor: it creates abstract points of view regarding the machine, causing it to be judged, not in its own right, but according to its costs and the results of its operation. Scientific knowledge, which sees in the technical object the practical application of a theoretical law, is not at the proper level of the technical domain either. Rather, it would appear that this task of understanding is left to the engineer of organization who would be like a sociologist or psychologist of machines, living in the midst of this society of technical beings as its responsible and inventive consciousness.

A genuine awareness of technical realities, grasped in their signification, corresponds to an open plurality of techniques. It cannot, moreover, be otherwise because a technical ensemble, even one that is not very extensive, comprises machines whose principles of operation are derived from very different scientific domains. So-called "technical" specialization most often corresponds to matters that are, strictly speaking, external to the technical objects (public relations, a particular form of commerce), rather than corresponding to a kind of operational schema within the technical objects; it is this specialization according to directives that are external to technics that creates the narrow-mindedness attributed to technicians by the cultivated man who intends to distinguish himself from them: it is a question of a narrow-mindedness of intentions, of ends, rather than a narrow-mindedness of information or of technical intuition. Today, it is rare for machines not to be, to some extent, simultaneously mechanical, thermal, and electric.

In order to restore to culture the truly general character it has lost, one must be capable of reintroducing an awareness of the nature of machines, of their mutual relations and of their relations with man, and of the values implied in these relations. This awareness requires the existence of a technologist or *mechanologist*, alongside the psychologist and the sociologist. Moreover, these fundamental schemas of causality and regulation that constitute an axiomatic of technology, must be taught in a universal fashion, in the same way the foundations of literary culture are taught. The initiation to technics must be placed on the same level as scientific education; it is as disinterested as the practice of the arts, and it dominates practical applications as much as theoretical physics does; it can attain the same degree of abstraction and symbolization. A child ought to know what self-regulation is, or what a positive reaction is, in the same way a child knows mathematical theorems.

This cultural reform, proceeding through expansion and not through destruction, could return the true regulative power that it has lost to today's culture. As the basis of significations, of means of expression, of justifications and of forms, a culture establishes regulatory communication among those who share that culture; arising from the life of the group, culture animates the gestures of those who ensure the command functions, by providing norms and schemas. However, before the great development of technics, culture incorporated the principal types of technics that give rise to lived experience, in the form of schemas, symbols, qualities, and analogies. Our current culture, by contrast, is the old culture, incorporating

as its dynamic schemas the state of artisanal and agricultural technics of bygone
16   centuries. And it is these schemas that serve as mediators between groups and
their leaders, imposing a fundamental distortion, as a result of their inadequacy to
technics. Power becomes literature, the art of opinion, advocacy of plausibility, and
rhetoric. The directive functions are false, because an adequate code of relations
between the governed reality and the beings who govern no longer exists: the governed reality comprises men and machines; the code merely relies on the experience
of the man working with tools, an experience which itself has become weakened
and remote because those who use this code haven't, like Cincinnatus, let go of
the handles of the plow only yesterday. The symbol is weakened into a mere lin-
guistic turn of phrase, reality is absent. A regulative relation of circular causality
cannot be established between the whole of governed reality and the function of
authority: information no longer achieves its purpose because the code has become
inadequate to the type of information it is supposed to transmit. Information
that will express the simultaneous and correlative existence of men and machines
must comprise the machines' schemas of functioning and the values they imply.
Culture, which has become specialized and impoverished, must once again become
general. By removing one of its principal sources of alienation and by re-establish-
ing regulative information, this extension of culture possesses political and social
value: it can give man the means for thinking his existence and situation according
to the reality that surrounds him. This work of broadening and deepening culture
also has a properly philosophical role to play because it leads to the critique of a
certain number of myths and stereotypes, such as that of the robot, or of perfect
automata at the service of a lazy and fulfilled humanity.

In order to raise this awareness, it is possible to attempt to define the technical
object itself, through the process of concretization and functional over-determi-
17   nation that gives it its consistency at the end-point of a process of evolution, thus
proving that it cannot be considered as a mere utensil.

The modalities of this genesis enable one to grasp the three levels of the technical
object and their non-dialectical temporal coordination: the element, the individual, and the ensemble.

Once the technical object is defined through its genesis, it becomes possible to
study the relations between the technical object and other realities, in particular
that of man at the stage of adulthood or childhood.

Lastly, considered as an object of value judgment, the technical object can provoke very different attitudes depending on whether it is considered at the level
of the element, at the level of the individual, or at the level of the ensemble.

At the level of the element, the process of its improvement does not introduce any upheavals that would engender anxiety by conflicting with acquired habits: this is the climate of eighteenth century optimism, which introduces the idea of continuous and indefinite progress, bringing about the constant improvement of man's lot. The technical individual entity, on the contrary, becomes for a certain time the adversary of man, his competitor, because man had centralized technical individuality within himself at a time when only tools existed; the machine thus takes the place of man because, as tool bearer, man used to do the job the machine now does. To this phase corresponds a dramatic and impassioned notion of progress, which turns into the rape of nature, the conquest of the world, and the exploitation of energies. This will to power expresses itself in the technophile and technocratic excesses of the thermodynamic era, which take on both a prophetic and cataclysmic spin. Finally, at the level of the technical ensembles of the twentieth century, this thermodynamic energeticism is replaced by information theory, whose content is normative and eminently regulative and stabilizing: the development of technics appears to be a guarantee of stability. The machine, as an element of the technical ensemble, becomes that which increases the quantity of information, increases negentropy, and opposes the degradation of energy: the machine, being a work of organization and information, is, like life itself and together with life, that which is opposed to disorder, to the leveling of all things tending to deprive the universe of the power of change. The machine is that through which man fights against the death of the universe; it slows down the degradation of energy, as life does, and becomes a stabilizer of the world.

This modification of the philosophical way of looking at the technical object announces the possibility of introducing the technical being into culture: this integration, which could not have taken place in a definitive way at the level of elements or at the level of individuals, will have a greater chance of stability at the level of ensembles; once technical reality has become regulative it can be integrated into culture, which is regulative in its essence. This integration could only have occurred by way of addition in the age when technicity resided in its elements, or by way of a breach and a revolution in the age when technicity resided in new technical individuals; today, technicity tends to reside in ensembles. For this reason, it can become a foundation for culture, to which it will bring a unifying and stabilizing power, making culture adequate to the reality which it expresses and regulates.

PART I
GENESIS AND EVOLUTION OF TECHNICAL OBJECTS

CHAPTER ONE

GENESIS OF THE TECHNICAL OBJECT: THE PROCESS OF CONCRETIZATION

*I. – The abstract technical object
and the concrete technical object*

Although the technical object is subject to genesis, it is difficult to define the gene- sis of each technical object, since the individuality of technical objects is modified throughout the course of this genesis; technical objects are not easily defined by attribution to a technical kind; it is easy to summarily distinguish kinds according to practical usage, as long as one accepts grasping the technical object according to its practical end; however, this is an illusory specificity, because no fixed structure corresponds to a definite usage. The same result may be obtained from very different functionalities and structures: a steam engine, a gasoline engine, a turbine, and an engine powered by springs or weights are all equally engines, but there is a more genuine analogy between a spring engine and a bow or a cross-bow than between the spring engine and a steam engine; the engine of a pendulum clock is analogous to a winch, while an electric clock is analogous to a door bell or a buzzer. Usage unites these heterogeneous structures and operations under the banner of genera and species that draw their signification from the relation between this functioning and another functioning, which is that of the human being involved in the action. That to which one thereby gives a single name — for instance the engine — can thus be multiple in one instance and may vary in time by changing its individuality.

However, instead of starting out with the individuality of the technical object, or even with its specificity, which is very unstable, it is preferable to reverse the problem, if we want to try to define the laws of its genesis in light of its individuality or specificity: one can define the individuality and specificity of the technical object on the basis of the criteria of its genesis: the individual technical object is not this or that thing, given *hic et nunc*, but that of which there is genesis.[1] The unity of the technical object, its individuality, and its specificity are the characteristics of consistency and convergence in its genesis. The genesis of the technical object partakes in its being. The technical object is that which is not anterior to its coming-into-being, but is present at each stage of its coming-into-being; the technical object in its oneness is a unit of coming-into-being. The gasoline engine is not this or that engine given in time and space, but the fact that there is a succession, a continuity that runs through the first engines to those we currently know and which are still evolving. As such, as in a phylogenetic lineage, a definite stage of evolution contains dynamic structures and schemas within itself that partake in the principal stages of an evolution of forms. The technical being evolves through convergence and self-adaptation; it unifies itself internally according to a principle of inner resonance. Today's automobile engine is the descendent of the engine from 1910 not simply because the engine of 1910 was built by our ancestors. Nor is today's automobile engine its descendant simply because it has a greater degree of perfection in relation to use; in fact, for some uses the engine from 1910 remains superior to an engine from 1956. For instance, it can tolerate extensive heating without galling or rod bearing failure, having been built with more flexibility and without fragile alloys such as Babbitt metal; it is more autonomous, due to its having a magnetic ignition. Old engines function reliably on fishing boats after having been taken from a disused automobile. It is through internal examination of the regimes of causality and forms, insofar as they are adapted to these regimes of causality, that the contemporary automobile engine is defined as posterior to the engine from 1910. In a contemporary engine each important item is so well connected to the others via reciprocal exchanges of energy that it cannot be anything other than what it is. The shape of the combustion chamber, the shape and size of the valves, and the shape of the piston all belong to the same system within which a multitude

1. According to determinate modalities that distinguish the genesis of the technical object from that of other types of objects: the aesthetic object, the living being. These specific modalities of genesis must be distinguished from a static specificity that one could establish after genesis by considering the characteristics of diverse types of objects; the point of using a genetic method is precisely to avoid using classification as a way of thinking that occurs after genesis only to distribute the totality of objects into genera and species suitable for discourse. The technical being retains the essence of its past evolution in the form of its technicity. According to the approach we shall call analectic, the technical being, as bearer of this technicity, can be the object of adequate knowledge only if the latter grasps the temporal sense of its evolution; this adequate knowledge is a culture of technics, distinct from technical knowledge, which is limited to the actuality of isolated schemas of operation. Considering that the relations that exist between one technical object and another at the level of technicity are horizontal as well as vertical, any form of knowledge that procedes by genera and species becomes inadequate: we will attempt to point out the way in which the relation between technical objects is transductive.

of reciprocal causalities exist. To such a shape of these elements corresponds a certain compression ratio, which in turn requires a determinate ignition timing; the shape of the cylinder head, as well as the metal it is made of, produce a certain temperature in the spark plug electrodes in relation to all the other elements of the   24 cycle; this temperature in turn causes a reaction leading to the characteristics of ignition and hence to the entire cycle. One could say that the contemporary engine is a concrete engine, whereas the old engine is an abstract engine. In the old engine each element intervenes at a certain moment in the cycle, and then is expected no longer to act upon the other elements; the pieces of the engine are like people who work together, each in their own turn, but who do not know one another.

Moreover, this is precisely how the functioning of thermal engines is explained to students in the classroom, each piece being isolated from the others like the lines that represent it on the blackboard in geometric space, *partes extra partes*. The old engine is a logical assemblage of elements defined by their complete and unique function. Each element can accomplish its own function best if it is, like a perfectly completed instrument, oriented entirely to accomplishing this function. A permanent exchange of energy between two elements appears as if it were an imperfection, unless this exchange itself belongs to the theoretical operation; furthermore there is a primitive form of the technical object, *the abstract form*, in which each theoretical and material unit is treated as an absolute, and is completed according to an intrinsic perfection that requires, in order for it to function, that it be constituted as a closed system; integration into an ensemble in this case raises a series of so-called technical problems that must be resolved and which are in fact problems of compatibility between already given ensembles.

These already given ensembles need to be maintained and preserved despite their reciprocal influences. What appears then are particular structures that one can call, for each constitutive unit, defense structures: the cylinder head of the thermal combustion engine bristles with cooling fins that are particularly well developed in the region of the valves, which is subject to intense thermal exchanges and high pressure. In the first engines these cooling fins are as if added from the outside to the theoretical cylinder and cylinder head, which are geometrically cylindrical; they   25 serve only one function, that of cooling. In more recent engines, these cooling fins also play a mechanical role, as ribs that resist the deformation of the cylinder head under the pressure of the gasses; in these conditions one can no longer distinguish the volumetric unit (cylinder, cylinder head) from the thermal dissipation unit; if, in an engine that uses ambient air for cooling, one were to remove the cylinder head's fins by sawing or grinding, then the volumetric unit constituted by the cylinder head alone would no longer be viable, even as a volumetric unit: it would be deformed under the gaseous pressure; the volumetric and mechanical unit has become coextensive with the unit of thermal dissipation because the structure of the ensemble is bivalent: the fins constitute a cooling surface of thermal exchanges with the stream of external air; these same fins, insofar as they are a part of the cylinder head, limit the size of the combustion chamber through their un-deformable contour, using less metal than would be required by a shell without ribs; the development of this unique structure is not a compromise, but a concomitance and a convergence: a ribbed cylinder head can afford to be thinner than a smooth cylinder head with the same rigidity; a thin cylinder head, in turn, allows for more efficient thermal exchanges than a thick cylinder head would allow; the bivalent fin-rib structure improves the cooling not only by increasing the thermal exchange area (which is what characterizes the fin as a fin), but also by permitting a thinning of the cylinder head (which is what characterizes the fin as ribbing).

26   The technical problem is thus one of the convergence of functions into a structural unit, rather than one of seeking a compromise between conflicting requirements. If, in the case just considered, a conflict subsists between two aspects of a single structure, then this is only because the position of the ribbing that would correspond to maximum rigidity is not necessarily the same as that which corresponds best to their fastest cooling by way of air flowing between the fins when the vehicle is running. In this case the builder might have to retain a mixed, incomplete aspect: the fin-ribbing, if positioned for optimal cooling, will have to be thicker and more rigid than if it were for ribbing alone. If, on the contrary, they are positioned perfectly to resolve the problem of obtaining rigidity, then their area is larger, in order to compensate the reduction of the thermal exchange that had been diminished because of the slowed airstream, via the development of a larger area; the very structure of the fins may in the end also be a compromise between two forms, requiring a greater development than if a single function were taken as the sole purpose of the structure. This divergence of functional directions is like a residue of abstraction within the technical object and it is the progressive reduction of this margin between the functions of plurivalent structures that defines the progress of a technical object; it is this convergence that specifies the technical object, because in any given epoch there is no infinite plurality of possible functional systems; there are far fewer technical species than there are usages to which technical objects are destined; human necessity is infinitely diversifiable, but the directions of convergence of technical species are finite in number.

The technical object thus exists as a specific type obtained at the end of a convergent series. This series goes from the abstract to the concrete mode: it tends toward a state which would turn the technical being into a system that is entirely coherent within itself and entirely unified.   27

II. – Conditions of technical evolution

What are the *reasons* for this convergence that manifests itself in the evolution of technical structures? A certain number of extrinsic causes no doubt exist, in particular those which tend to produce the standardization of spare parts and organs. Nevertheless, these extrinsic causes are not more powerful than those that tend toward the multiplication of types, appropriated for an infinite variety of needs. If technical objects do evolve toward a small number of specific types then this is by virtue of an internal necessity and not as a consequence of economic influences or practical requirements; it is not the production-line that produces standardization, but rather intrinsic standardization that allows for the production-line to exist. An effort to discover the reason for the formation of specific types of technical objects within the transition from artisanal production to industrial production would mistake the consequence for its condition; the industrialization of production is rendered possible by the formation of stable types. Artisanal production corresponds to the primitive stage of the evolution of the technical object, i.e., to the abstract stage; industry corresponds to the concrete stage. The *made-to-measure* aspect one finds in the product of artisanal work is inessential; it is the result of this other, essential aspect of the abstract technical object: namely, that it is grounded in an analytical organization that always leaves the path open for new possibilities; these possibilities are the external manifestation of an internal contingency. In the confrontation between the coherence of technical work and the coherence of a system of the needs of utilization, it is the coherence of utilization that prevails, because the technical object that is made to measure is in fact an object without intrinsic measure; its norms are derived from the outside: it has not yet realized its   28
internal coherence; it is not a system of the necessary; it corresponds to an open system of requirements.

Conversely, during the industrial stage, the object achieves its coherence and it is the system of needs that is now less coherent than the system of the object; needs mold themselves onto the industrial technical object, which in turn acquires the power to shape a civilization. It is utilization that becomes an ensemble chiseled to the measures of the technical object. When individual fancy calls for a customized automobile, the manufacturer can do no more than take a serial engine, a serial chassis, and externally modify some aspects, adding decorative details or externally adjusted accessories to the automobile, which is really the essential technical object: what can be made to measure are inessential aspects, because they are contingent.

The type of relation that exists between these inessential aspects and the nature proper to the technical type is a negative one: the more the car is required to answer to a large number of user demands, the more its essential characteristics are encumbered with external servitude; the bodywork burdens itself with accessories, shapes no longer correspond to the structures facilitating the best air flow. The made-to-measure aspect is not only inessential, it goes against the essence of the technical being, it is like a dead weight imposed from the outside. The car's center of gravity rises, its mass increases.

It is not enough, however, to claim that the evolution of the technical object occurs via a passage from an analytic order to a synthetic order, conditioning the passage from artisanal production to industrial production: even if this evolution is necessary, it is not automatic and one ought to seek the causes of this evolving movement. These causes essentially reside in the imperfection of the abstract technical object. Because of its analytic aspect, this object uses more material and requires more construction work; it is logically simpler, yet technically more complicated, because it is made up of a convergence of several complete systems. It is more fragile than the concrete technical object, because the relative isolation of each system that constitutes a functional sub-system threatens, in case of its malfunction, the preservation of the other systems. In an internal combustion engine the cooling process might thus be accomplished by an entirely autonomous sub-system; if this sub-system ceases to work, the engine might deteriorate; if, on the contrary, the cooling process is the effect resulting from the solidarity of the functioning of the whole, then the functioning itself implies cooling; in this sense, an air-cooled engine is more concrete than a water-cooled engine: thermal infrared radiation and convection are effects that cannot but take place; they are necessitated by its functioning; cooling by water is semi-concrete: if it were accomplished entirely by a thermosiphon,* it would be almost as concrete as cooling by air; but the use of a water pump, receiving energy from the engine via the transmission belt, increases the element of abstraction of this kind of cooling; one could say that cooling by water is concrete in terms of a safety system (the presence of water enables summary cooling for a few minutes when the transmission from engine to pump is deficient, thanks to the absorption of calorific energy through evaporation); in its normal functioning, however, it is an abstract system; moreover, an

---

*   cf. glossary of technical terms. [TN]

element of abstraction still subsists in the possibility of the cooling circuit lacking water. Ignition via an ignition coil and accumulator battery is, likewise, more abstract than ignition by magneto,* which is itself more abstract than ignition by the compression of air followed by fuel injection, such as those used in diesel engines. One could say that in this sense an engine with a magnetic fly-wheel and which is air-cooled is more concrete than a typical car engine; each piece plays several roles here; it is not surprising that the scooter is the brain-child of an engineer specializing in aviation; while the automobile can afford to preserve remnants of abstraction (cooling by water, ignition by battery and coil), aviation is obliged to produce the most concrete technical objects, in order to increase safe functioning and diminish dead weight. 30

Thus properly speaking, there is a convergence of economic constraints (a diminished quantity of raw material, of work and of energy consumption during use) and technical requirements: the object cannot be self-destructive, it must maintain itself in a stable state of functioning for as long as possible. As far as these two types of cause — the economic and the properly technical — are concerned, it would appear that it is the latter that predominates in technical evolution; economic causes indeed exist in all domains; yet it is mostly within the domains where technical constraints prevail over economic constraints (aviation, military equipment) that become the most active sites for progress. Indeed, economic causes are not pure; they interfere with a diffuse network of motivations and preferences that attenuate or even reverse them (a taste for luxury, the user's desire for very apparent novelty, commercial propaganda), to such an extent that in domains where the technical object is known through social myths or fads in public opinion, rather than being appreciated in itself, certain tendencies toward complication come to light; some car manufacturers thus present the use of overabundant automatism in accessories or the systematic recourse to servo-mechanisms as an increase in perfection, even where direct command does not in the least exceed the physical 31 strength of the driver: some even go so far as to find a sales argument and proof of superiority in the suppression of direct means, as for instance that of the use of the crank as a back-up means of starting the engine, which in fact consists in making its operation more analytic in subordinating it to the use of available electric energy accumulated in batteries; technically this represents a complication, whereas the manufacturer presents this suppression as a simplification that would show how modern the car is, thereby making the unpleasant affective connotations of the stereotypical image of a car engine that is difficult to start a thing of the past. A nuance of ridicule is thus projected onto other cars — those that preserve the crank — which are somehow out of date, discarded into the past through an artifice of presentation. The automobile, a technical object charged with psychic and social inferences, is not suitable for technical progress: the automobile's progress comes from neighboring domains, such as aviation, shipping, and transportation trucks.

The specific evolution of technical objects occurs neither in an absolutely continuous nor completely discontinuous manner; it is made up of stages that are defined by the fact that they produce successive systems of coherence; between stages marking a structural re-organization there can be an evolution of a continuous kind; this is due to the progressive perfection of details resulting from experience and use, and from the production of better adapted raw materials or auxiliary devices; for thirty years the automobile engine improved through the use of metals that were better adapted to the conditions of utilization, through the increase of the compression ratio as a result of research into fuels, and through the study of the particular form of cylinder heads and piston heads in relation to the phenomenon of detonation. *

32   The problem that consists in producing combustion while avoiding detonation can be resolved only through work of a scientific kind on the propagation of the explosive wave at the heart of a carburized mix, at different pressure levels, at different temperatures, with diverse volumes and starting from determinate ignition points. This effort, however, does not itself lead directly to applications: the experimental work remains to be accomplished and there is a technicity proper to this path toward progressive perfection. What is essential in the coming-into-being of this object are the structural reforms that facilitate the technical object's self-specification; even if the sciences were to stop progressing for a time, the progress of the technical object toward specificity would continue; the principle of this progress is effectively the manner in which the object causes and conditions itself in its functioning and in the reactions of its functioning on its utilization; the technical object, issued forth from the abstract work of the organization of sub-systems, is the theater of a certain number of reciprocal causal relations.

It is due to these relations, given certain limits of the conditions of utilization, that the object encounters obstacles within its own operation: *the play of limits, whose overcoming constitutes progress, resides in the incompatibilities that arise from the progressive saturation of the system of sub-ensembles;*[2] yet because of its very nature, this overcoming can occur only as a leap, as a modification of the internal distribution of functions, a rearrangement of their system; what was once an obstacle must become the means of realization. Such is the case regarding the evolution of

---

2.   They are the conditions of a system's individuation.

the electronic tube, whose most common type is the radio lamp. What caused the structural reforms, whose end point is today's series of lamps, were the internal obstacles preventing the proper functioning of the triode. One of the most prob-   33
lematic phenomena of the triode was the significant reciprocal capacitance within the system formed by the command grid and anode; this capacitance was in fact the capacitive coupling between two electrodes, and one could not notably increase the size of these electrodes without running the risk of initiating self-oscillation; this inevitable internal coupling had to be compensated for by way of an external assembly, in particular through neutrodyning, which was achieved by using an assembly of symmetrical lamps with anode-grid coupling.

In order to resolve the difficulty rather than avoid it, an electrostatic plate was inserted into the triode, between the command grid and the anode; this addition, however, adds more than just the advantage of an electric screen. The screen cannot exclusively accomplish the function of decoupling for which it was meant: placed inside the space between grid and anode, it intervenes with its differential potential (with respect to both grid and anode) as another grid with respect to the anode and as anode with respect to the grid. It needs to be brought to a potential superior to that of the grid and inferior to that of the anode; failing this condition, no electron passes through or else the electrons arrive at the screen rather than the anode. The screen thus acts upon the electrons in transit between the grid and the anode; it is itself both a grid and an anode; these two conjugated functions are not obtained intentionally; they impose themselves by themselves as a surplus that results from the systemic aspect of the technical object. In order to introduce the screen into the triode without disturbing its operation it must at once fulfill functions relating to the electrons in transit, as well as its electrostatic function. Considered as a simple electrostatic plate, it could be brought to any degree of voltage, provided this remains a direct current; but it would thereby upset the dynamic functioning of the triode. It necessarily becomes an accelerating grid for the flow of electrons and also plays a positive role in the dynamic functioning: it notably increases inter-   34
nal resistance, and consequently the coefficient of amplification, when brought to a determinate level of voltage, defined by its exact position in the grid-anode space. The tetrode is then no longer simply a triode without electrostatic coupling between anode and command grid; the tetrode is an electronic tube with great gain, with which one can obtain an amplification of the order of 200 rather than an amplification of 30 to 50 as far as the triode is concerned.

However, this discovery also contained an inconvenience: what became problematic in the tetrode was a phenomenon of the secondary emission of electrons by the anode, which tended to bounce all the electrons coming from the cathode (the primary electrons) and which had already passed through the command grid back onto the screen in the opposite direction; Tellegen therefore introduced another screen between the first screen and the anode: once brought to a negative potential with respect to the anode and screen (generally the potential of the cathode or even more negative a potential) this wide-meshed grid no longer obstructs the arrival of accelerated electrons going from the cathode to the anode, but acts as a negatively polarized command grid and stops the return of secondary electrons going against the direction of the flow. The pentode is thus the outcome of the tetrode, in the sense that it contains a supplementary command grid with a fixed potential that completes the dynamic schema of functioning; the same effect of irreversibility, however, can be obtained by concentrating the flow of electrons into beams; if the bars of the accelerating screen-grid are placed in the electric shadow of the command grid wires, then the phenomenon of secondary emission is greatly reduced. Furthermore, the capacitance variation between cathode and screen-grid during functioning becomes very minor (0.2 pF instead of 1.8 pF), practically
35 preventing all frequency drift when the tube is used in a circuit that generates oscillations. Consequently, one could say that the tetrode's schema of functioning is not perfectly complete in itself, if one considers the screen merely as an electrostatic shield, that is, as a barrier that can be brought to any continuous voltage whatsoever; this definition, however, is too broad; the definition of the tetrode needs to include the multifunctionality of the screen into the electronic tube, which happens by reducing both the margin of indeterminacy of the continuous voltage to be applied to the screen (in order for it to be an accelerator) and that of its position in the grid-anode space; a first reduction consists in specifying that the continuous voltage will have to be an intermediary between the grid and the anode's voltage; one thus obtains a structure that is stable with respect to the acceleration of primary electrons, but which still remains indeterminate with respect to the transit of secondary electrons coming from the anode; this structure in turn is still too open, too abstract; it can be closed, so that it corresponds to a necessary and stable functioning, either via a supplementary structure — the suppressive grid or third grid — or through greater precision in the positioning of the screen-grid with respect to the other elements, which consists in aligning its wires with those of the command grid. It is worth noting that the addition of a third grid is equivalent to an increase in the degree of determinacy in the screen-grid's positioning: there

is a reversibility between the functional aspect of the determination of structures already existing by their reciprocal causality and the functional aspect of a supplementary structure; closing down the system of the existing structures' reciprocal causality with increased determinacy is equivalent to adding a new structure specialized in accomplishing a determinate function. In the technical object there is a reversibility between function and structure; over-determination of the system of structures within the regime of their functioning makes the technical object more concrete by stabilizing its functioning without adding a new structure. A tetrode with targeted beams is equivalent to a pentode; it is even superior in its function as a power amplifier of acoustic frequencies, because of the lower degree of distortion it produces. The adjunction of a supplementary structure only constitutes genuine progress for the technical object if this structure incorporates itself concretely into the totality [*ensemble*] of dynamical schemas of functioning; for this reason we will say that the tetrode with targeted beams is more concrete than the pentode.

One must not confuse the increase of the concrete aspect of the technical object with the increase of the technical object's possibilities via a complication of its structure; for instance, a twin-grid lamp (which allows separate action on two independent control grids in a single cathode-anode space) is no more concrete than a triode; it is of the same order as the triode, and could, if need be,[3] be replaced by two independent triodes, whose cathodes and anodes would be connected externally while leaving the command grids independent. Conversely, the tetrode with targeted beams, is more evolved than Lee de Forest's triode, because it is the realization of a development, the refinement of a primitive schema of the flow of the electrons' modulation by fixed or variable electric fields.

The primitive triode has a greater degree of indeterminacy than modern electronic tubes, because the interactions between structural elements during the course of its functioning are not defined, except for a single one among them, namely the modulating function of the electric field created by the command grid. The inconveniences that emerge of their own accord during functioning are *transformed into stable functions* by the successive specifications and closures of the system: the necessity for the grid's negative polarization to counteract heating and secondary emissions also contains the possibility of splitting the primitive grid's function into that of both a command grid and an accelerating grid; in a tube with an accelerating grid, the command grid's negative polarization can be reduced to a few volts, and in some cases one volt; the command grid becomes almost exclusively a command grid: its function is more efficient and the tube's gain increases.

36

37

---

3. Not entirely because each grid can modulate all the way, whereas with two lamps, it would be halfway.

The command grid is brought closer to the cathode; conversely, the second grid, the screen, moves farther away and establishes itself roughly at an equidistance between the anode and cathode. At the same time functioning becomes stricter; the dynamic system closes in on itself just as an axiomatic saturates. The first triode's gain could be regulated through the potentiometric variation of a cathode's heating voltage, acting on the density of electron flow; this possibility is barely available anymore in pentodes with a large gain, whose characteristics would be profoundly altered by an appreciable variation in the heating voltage supply.

It seems contradictory, of course, to affirm that the evolution of the technical object obeys both a process of differentiation (the triode's command grid is divided into three grids within the pentode) and a process of concretization, where each structural element fulfills several functions rather than a single one; but these two processes are in fact tied to each other; differentiation is possible because, in a conscious and calculated manner and in view of a necessary result, it enables the integration of correlative effects of the global functioning into the functioning of the ensemble, effects which had until then been more or less corrected by palliatives that were separate from the principle function.

38 The same type of evolution is noticeable in the passage from the Crookes tube to the Coolidge tube; the first is not only less efficient than the second; it is also less stable in its functioning, and more complex; the Crookes tube in fact uses the cathode-anode voltage to disassociate molecules or atoms of mono-atomic gases into positive ions and electrons, in order to then accelerate these electrons, conferring upon them an important amount of kinetic energy prior to their collision with the anticathode; conversely, in a Coolidge tube, the function of producing electrons is dissociated from that of the acceleration of already produced electrons; production is achieved through a thermoelectric effect (improperly called thermionic, probably because it replaces the production of electrons through ionization), and subsequently acceleration takes place; the functions are thus purified by their dissociation, and the corresponding structures are both more distinct and richer; the hot cathode of a Coolidge tube is richer from the point of view of its structure and its function than the cold cathode in a Crookes tube; and yet it is also a perfect cathode from the electrostatic point of view; all the more so since it has a rather narrow localization of the source of thermo-electrons, and the form of the cathode's surface surrounding the filament determines an electrostatic gradient that enables focalization of the electrons in a narrow beam falling onto the anode (a few square millimeters in current tubes); a Crookes tube, on the contrary, does not have a defined location narrow enough for the source of electrons that would

enable it to focalize a beam efficiently and thus obtain a source of X-rays nearing the ideal punctuality.

Moreover, the presence of ionizable gas in a Crookes tube not only had the inconvenience of being unstable (the hardening of the tube from a fixation of molecules on the electrodes; the necessity of contriving valves in order to reintroduce gas into the tube); the presence of the gas also brought with it another essential inconvenience: the gas molecules became an obstacle for electrons that had already been produced and which were in the process of acceleration within the electrical field between the cathode and the anode; this inconvenience offers a typical example of a functional antagonism within the processes of an abstract technical object's functioning: the very gas that is necessary to produce the electrons that are to be accelerated becomes an obstacle to their acceleration. This antagonism disappears with the Coolidge tube, which is a deep vacuum tube. It disappears because the groups of synergetic functions are allocated to specific structures; each structure attains a greater functional wealth from this redistribution as well as a more perfect structural precision; this is the case for the cathode, which, rather than being a simple spherical cap or hemisphere made of any kind of metal, becomes an ensemble consisting of a parabolic bowl at the heart of which there is a filament producing thermo-electrons; the anode, which in the Crookes tube occupied an indifferent position with respect to the cathode, converges geometrically with the previous anticathode; the new anode-anticathode plays the two synergetic roles: that of the production of a potential difference with respect to the cathode (the role of the anode) and that of the obstacle against which the electrons, accelerated by differential potential, collide, thus transforming their kinetic energy into luminous energy of a very short wavelength.

These two functions are synergetic because it is only after having sustained the entirety of the potential drop of the electric field that the electrons acquire maximum kinetic energy; it is thus both at this moment and in this location that it becomes possible to extract the greatest electromagnetic energy by stopping them abruptly. The new anode-cathode finally plays a role in the evacuation of heat produced (as a result of the poor efficiency of the transformation of the electron's kinetic energy into electromagnetic energy, roughly 1%) and this new function is fulfilled in perfect accord with the two previous ones: a plate of hard-to-fuse metal, such as tungsten, is embedded into the bevel-sawed solid copper bar that forms the anode-cathode at the point of impact of the electron beam; the heat that develops on this plate is conducted outside the tube via the copper bar, which extends to the outside as cooling fins.

There is a synergy between the three functions, as the electric characteristics of the highly conductive copper bar goes hand in hand with the thermal characteristics of this same bar as a heat conductor; furthermore, the beveled section of the copper bar satisfies the function of a target-obstacle (anticathode), the acceleration of electrons (anode) and the evacuation of produced heat. One could say that, under these conditions, the Coolidge tube is a Crookes tube that is both simplified and concretized, in which each structure fulfills a greater but synergetic number of functions. The imperfection of the Crookes tube, its abstract and artisanal aspect, requiring frequent touch-ups in its functioning, came from the antagonism of functions fulfilled by the rarefied gas; it is this gas which is eliminated in the Coolidge tube. The fuzzy, indefinite structure corresponding to the ionization is entirely replaced by the new thermoelectric aspect of the cathode, which is perfectly clear and quantitatively adjustable.

These two examples tend to show that differentiation goes in the same direction as the condensation of multiple functions within the same structure, because the differentiation of structures within a system of reciprocal causalities allows one to suppress side-effects that were hitherto obstacles (by integrating them into the functioning). The specialization of each structure is a specialization of a synthetic positive functional unit, freed from undesired side-effects that affect functioning; the technical object progresses by way of an internal redistribution of functions into compatible units, replacing the contingency or antagonism of the primitive distribution; specialization does not occur *function after function*, but *synergy after synergy*; it is the synergetic group of functions and not the unique function that constitutes the true sub-system in the technical object. It is because of this search for synergies that the technical object's concretization can translate into an element of simplification; the concrete technical object is one that is no longer in conflict with itself, one in which no side-effect is detrimental to the functioning of the ensemble or left out of this functioning. In this manner and for this reason a function can be fulfilled by several synergistically associated structures in the technical object that has become concrete, whereas in the primitive and abstract technical object each structure is charged with the accomplishment of a definite function, and generally only one. The essence of the technical object's concretization is the organization of functional sub-ensembles within the total functioning; on the basis of this principle one can understand in what sense the redistribution of functions occurs in the network of different structures, both in the abstract technical object and in the concrete technical object: each structure fulfills several functions; but in the abstract technical object, it only fulfills one essential and positive function, integrated into the functioning of the ensemble; in the concrete technical object, all the functions fulfilled by the structure are positive, essential, and integrated into the functioning of the whole; the marginal consequences of the functioning, eliminated or attenuated in the abstract technical object by corrective measures, become stages or positive aspects in the concrete object; the schema of functioning incorporates marginal aspects; consequences that were irrelevant or harmful become chain-links in its functioning.

This progress presupposes that the engineer consciously endows each structure   42
with characteristics that correspond to all the components of its functioning, as if there were no difference between the artificial object and a physical system studied from the point of view of all knowable aspects of exchanges of energy, as well as physical and chemical transformations; each piece, in the concrete object, is no longer simply that which essentially corresponds to the accomplishment of a function desired by the builder, but part of a system where a multitude of forces act and produce effects that are independent of the fabricating intention. The concrete technical object is a physico-chemical system in which reciprocal actions take place according to all the laws of the sciences. The objective of the technical intention can attain perfection in the construction of an object only if it becomes identical to universal scientific knowledge. It should be emphasized that this latter knowledge must indeed be universal, because the fact that the technical object belongs to the class of fabricated objects, answering to this particular human need, does not in turn limit and in no way defines the type of physico-chemical actions that can occur in this object or between this object and the outside world. The difference between the technical object and the physico-chemical system studied as an object only resides within the imperfection of the sciences; the scientific knowledge that serves as a guide to predicting the universality of reciprocal actions exerted within the technical system is still affected by a certain imperfection; it doesn't allow for an absolute prediction with rigorous precision of all effects; this is why a certain distance remains between the system of technical intentions corresponding to a defined objective and the scientific system of knowledge of causal interactions that realize this objective; the technical object is never fully known; for this very reason, it is never completely concrete, unless it happens through a rare chance occur-   43
rence. The ultimate allocation of functions to structures and the exact calculation of structures could only be accomplished if the scientific knowledge of all phenomena likely to exist in the technical object were completely acquired; since this is not the case, a certain difference subsists between the technical scheme of the object (containing the representation of a human objective) and the scientific picture

of phenomena for which it is the base (containing only schemas of reciprocal or recurrent efficient causality).

The concretization of technical objects is conditioned by way of narrowing the interval that separates the sciences and technology; the primitive artisanal phase is characterized by a weak correlation between the sciences and technology, whereas the industrial phase is characterized by a strong correlation. The construction of a determinate technical object can only become industrial when this object has become concrete, which means that it is known in an almost identical manner according to the intention of construction and according to the scientific view. This explains the fact that certain objects could be manufactured in an industrial manner well before others, a winch, a hoist, snatch blocks, and a hydraulic press are technical objects in which, for the most part, the phenomena of friction, electrical charging, electrodynamic induction, thermal and chemical exchanges can be neglected without leading to the destruction or malfunction of the object; rational classical mechanics are sufficient for a scientific knowledge of the principal phenomena that characterize the functioning of these objects we call simple machines: however, it would have been impossible to industrially manufacture a centrifugal gas pump or a thermal engine in the seventeenth century. The first industrially produced thermal engine, which was the Newcomen atmospheric engine, simply used the process of depression, because the phenomenon of the condensation of steam under the influence of cooling was scientifically known. Electrostatic machines also remained artisanal nearly to the present day, because the phenomena of the production and transport of charges via dielectrics and then flowing of charges via the Corona effect, which had been qualitatively known since at least the eighteenth century, had not yet been subjected to rigorous scientific study; after the Wimshurst machine, even the Van de Graaff generator retained something artisanal, despite its large size and greater power.

III. – The rhythm of technical progress; continuous and minor improvements; discontinuous and major improvements

It is thus essentially the discovery of functional synergies that characterizes progress in the development of the technical object. It is appropriate to ask, then, whether this discovery takes place all at once or in a continuous manner. In terms of the reorganization of structures affecting functioning, it happens abruptly, but can contain several successive stages; the Coolidge tube, for instance, could not have been conceived of before Fleming's discovery of the production of electrons by a heated metal; and yet Coolidge's static anode-anticathode tube is not necessarily the last version of a tube that produces X-rays or gamma rays. It can be improved and appropriated for more specific uses. An important refinement, for instance, enabling the acquisition of a source of X-rays closer to the ideal geometric point, consisted in employing a solid anode plate, mounted on an axis within the tube: this plate can be set in motion by a magnetic field created by an inductor placed outside the tube and in relation to which the plate is a rotor containing an armature; the region of impact of electrons becomes a circular line close to the edge   45 of the copper plate, and thus offers ample possibilities for thermal dissipation; however, the place where impact takes place is fixed, in a static and geometric manner, with respect to the cathode and the tube: the beam of X-rays thus comes from a geometrically fixed source, even though the anticathode rotates at high speed within this fixed point. Tubes with a rotating anode allow for an increase in power without increasing the size of the region of impact, or for a reduction in the size of the region of impact without diminishing the power; yet this rotating anode fulfills the functions of acceleration and absorption of electrons as perfectly as a fixed anode; it better fulfills the function of the evacuation of heat, which allows for an improvement of optical characteristics of the tube for a determinate degree of power.

Should one therefore consider the invention of the rotating anode to be a structural concretization of the Coolidge tube? — No, because it mostly plays the role of diminishing an inconvenience that couldn't be converted into a positive effect of the overall functioning [*fonctionnement d'ensemble*]. The inconvenience of the Coolidge tube, the residual aspect of antagonism that subsists in its functioning, is the poor yield of its conversion of kinetic energy into electromagnetic radiation; this poor yield probably does not constitute a direct antagonism between functions, but it does practically convert into a real antagonism; if the tungsten plate and the copper bar's melting point were infinitely high, then one could achieve the concentration of a very fine, powerful beam of very rapid electrons; but as the melting temperature for tungsten is in fact attained very quickly, one is limited by this poor yield producing a large quantity of heat, and one must resign oneself to sacrificing either the concentration of the beam, the density of electron flow, or the speed of electrons, which means sacrificing the punctuality of the X-rays' source, the quantity of radiated electromagnetic energy or the penetration of produced X-rays. If it were possible to discover a way of increasing the yield of energy trans-   46 formation that takes place on the anticathode piece, all the characteristics of the

Coolidge tube would be improved, thereby removing or diminishing the greatest antagonisms that subsist in this way of functioning. (A much weaker antagonism consists in the fact that the beam cannot be rigorously concentrated, because of the mutual repulsion of electrons, since they are affected by an electric charge with the same sign; this could be compensated for by way of devices aimed at a concentration comparable to those of either the cathode-ray oscilloscopes, or the electrostatic or electromagnetic lenses of electronic microscopes.) The rotating anode allows for a reduction of the consequences of the antagonism between precision and power, between optical and electronic characteristics.

There are thus two types of refinement: those which modify the distribution of functions, increasing the synergy of functioning in an essential way, and those which, without modifying this distribution, diminish the nocuous consequences of residual antagonisms; a more regular system of lubrication in the engine, the use of self-lubricating bearings, this order of minor improvements includes the use of more resistant metals or of more solid assemblages. The discovery of the high emission power of certain oxides or metals, such as thorium, has thus made it possible to build oxide cathodes that work at lower temperatures and absorb less heating energy for the same density of electron flux in the electronic tube. However important this refinement may be in practice, it nevertheless remains minor, and is suitable only for certain types of electronic tubes because of the fragility of the oxide coating. The rotating anode of Coolidge's high power tube is another minor refinement; it provisionally replaces a major improvement which would consist in discovering high yield energy transformation, thus enabling a reduction of the power deployed to accelerate electrons to a few hundred watts, which in radiography tubes is currently on the order of several kilowatts.

In this sense, one can say that minor improvements obstruct major improvements, because they may mask the technical object's true imperfections by compensating for true antagonisms with an inessential artifice that is incompletely integrated into the functioning of the ensemble; the danger of abstraction recurs once again at the level of minor improvements; for instance, the Coolidge tube with a rotating anode is less concrete than a tube with static cooling facilitated by a copper bar and air cooling fins; if, for whatever reason, the rotation of the anode stops during the tube's functioning, the point receiving the concentrated beam of electrons in the anode will almost instantaneously melt and the whole tube would be damaged; this analytical aspect of functioning thus requires new kinds of adjustment, a safety system based on the conditioning of one functioning by means of another functioning; in the analyzed case, the generator of anodic voltage must be capable of functioning only if the anode is already rotating; power to the transformer that supplies the anode voltage is controlled by the passage of current to the inductor for the anode motor, through a relay; however, this subordination does not completely reduce the analytic distance introduced by the rotating anode device; current may pass into the inductor while the anode may not be effectively rotating, due, for instance, to a weakening of its axis; the transmitter may also stay switched on even if the inductor is not live.

An extreme complication and refinement of ancillary safety or compensation systems can only tend toward an equivalence with the concrete technical object without attaining or even preparing for it, because the path chosen is not that of concretization. The path of minor improvements is one of detours, which may be useful in some cases for practical use, but they hardly make the technical object evolve. By dissimulating the true schematic essence of each technical object behind a pile of complex palliatives, minor improvements entertain a false consciousness of a continuous progress of technical objects, diminishing the value and feeling of urgency for essential transformations. For this reason, continuous minor improvements present no clear-cut frontier with respect to this false novelty that commerce demands in order to present a more recent object as being superior to older ones. Minor improvements can be so nonessential as to allow for the cyclical rhythm of fashionable forms to superimpose itself over the essential shapes of use objects. 48

Therefore, it is not enough to say that the technical object is that for which there is a specific genesis proceeding from the abstract to the concrete; the point has to be made that this genesis occurs because of essential, discontinuous improvements, as a result of which the internal schema of the technical object is modified in leaps rather than following a continuous line. This does not mean that the development of the technical object happens by chance and outside any assignable meaning; on the contrary, the minor improvements are, to a certain extent, those which occur by chance, overcharging the pure shapes of the essential technical object with their uncoordinated proliferation. The true stages of the technical object's improvement occur through mutations, but through mutations which are oriented: the Crookes tube contains the potential for the Coolidge tube, since the intention that organizes and stabilizes itself in the Coolidge tube by purifying itself, pre-existed in the Crookes tube, in a confused yet real state. A great many abandoned technical objects are unfinished inventions that remain as an open virtuality and could be taken up again, prolonged in another domain, according to their deep intention, their technical essence. 49

IV. – Absolute origins of the technical lineage

As with any evolution, that of technical objects poses the problem of its absolute origins: what is the first term one can attribute to the birth of a specific technical reality? Before the pentode and tetrode there was Lee de Forest's triode; before Lee de Forest's triode there was the diode. But what was there before the diode? Is the diode an absolute origin? Not entirely; thermoelectric emission, of course, was unknown, but the phenomena of the transfer of charges in space via an electric field had been known for a long time: electrolysis had been known for a century, and the ionization of gases for several decades; thermionic emission is necessary for the diode as a technical schema, because the diode wouldn't be a diode if there were reversibility of the transfer of electric charges; this reversibility doesn't exist in normal conditions, because one of the electrodes is hot, and therefore emissive, and the other cold, and therefore non-emissive; what makes a diode essentially a diode, a two way valve, is the fact that the hot electrode can be either cathode or anode almost indifferently, whereas the cold electrode can only be an anode since it cannot emit electrons; it can only attract, if it is positive, but not emit, even if it is negative with respect to another electrode. If one applies external voltage to the electrodes, a current will flow as a result of the thermoelectronic effect if the cathode is negative with respect to the anode, whereas no current will flow if the
50 hot electrode is positive with respect to the cold electrode. It is this discovery of a condition of functional dissymmetry between the electrodes that constitutes the diode, and not, properly speaking, that of a transfer of electrical charges through a vacuum mediated by an electric field: experiments with ionization of monoatomic gases had already shown that free electrons can move in an electric field; but this phenomenon is reversible, not polarized; if the tube of purified gas is turned upside down, then the positive column and luminous rings change sides in relation to the tube, but remain on the same side in relation to the direction of current coming from the generator. The diode is made up of the association of this reversible phenomenon of the transfer of electric charges by a field and the condition of irreversibility created by the fact that the production of transferable electric charges is the production of a single kind of (only negative) electric charges and by only one of the two electrodes, the hot electrode; the diode is a vacuum tube in which there is a hot electrode and a cold electrode, between which an electric field is created. What we have here is indeed an *absolute beginning*, residing in the association of this condition of irreversibility of the electrodes and of this phenomenon of

transfer of electric charges through a vacuum: it is a *technical essence* that is created. The diode is an asymmetrical conductance.

Nevertheless one should note that this essence is larger than the definition of the Fleming valve; several other procedures have been discovered for the creation of asymmetrical conductance; the contact between galena and metal, between copper and copper oxide, between selenium and another metal, between germanium and a tungsten tip, and between crystallized silicon and a metal tip are asymmetrical conductances. In the end, a single photoelectric cell can be considered a diode, since the photoelectrons behave like thermoelectrons in the vacuum of a cell (in the case of a vacuum cell, and also in that of a gas cell, but the phenomenon is com- 51 plicated by the emission of secondary electrons in addition to the photoelectrons). Should the name *diode* then be reserved for the Fleming valve? Technically, the Fleming valve can be replaced in several applications, by germanium diodes (for low intensities with high frequencies) or by a selenium or cupric oxide rectifier, for applications with low frequency and high intensity. Usage, however, does not offer good criteria: one could also replace the Fleming valve with a rotating converter,* which is a technique whose essential schema is entirely different from that of the diode. As a matter of fact, the thermoelectric diode constitutes a definite genus, which has its own historical existence; beyond this genus there is a *pure schema of functioning* that is transposable to other structures, for instance that of imperfect or semi-conductors; the schema of functioning is the same to such an extent that one can indicate a diode on a theoretical schematization with a sign (asymmetrical conductance: $\overset{\text{Anode}}{\scriptstyle(+)}$ ——▷|—— $\overset{\text{Cathode}}{\scriptstyle(-)}$) that does not predetermine the type of diode employed, leaving a freedom of choice to the manufacturer. However, the pure technical schema defines a type of existence of the technical object, grasped in its ideal function, which is different from the reality of a historic type; historically, the Fleming diode is closer to Lee de Forest's triode than to a germanium rectifier, to cupric oxide, or selenium and iron, which are nonetheless signaled by the same schematic symbols and in some cases fulfill the same functions to the point of being substitutable for the Fleming diode. This is because the whole essence of the Fleming valve is not reducible to its aspect of asymmetrical conductance; it is also that which produces and transfers a flow of electrons that can be slowed down, accelerated or deviated, as well as dispersed or concentrated, repulsed or attracted; the technical object exists not only as a result of its functioning within external 52 devices (asymmetrical conductance), but through phenomena for which it is in itself the basis: this is where its *fecundity* comes from, a *non-saturation* giving it posterity.

The primitive technical object can be considered a non-saturated system: any ulterior improvements it receives intervene as progress of the system toward saturation; viewed from the outside, one could think that the technical object alters and changes its structure rather than improving itself. But one could say that the technical object evolves by generating a family: the primitive object is the ancestor of this family. Such evolution could be called a *natural technical evolution.* In this sense, the gas engine is an ancestor of the gasoline engine and the diesel engine; the Crookes tube is the ancestor of the Coolidge tube; the diode is the ancestor of the triode and other tubes with multiple electrodes.

At the origin of each of these series, there is a definite act of invention; the gas engine follows, in a sense, from the steam engine; the disposition of its cylinder, of its piston, of its system of transmission, of its distribution by valves and ports is analogous to that of the steam engine; but it comes about by way of the steam engine like the diode comes about by way of the discharge tube through ionization in gases: what was needed was also a new phenomenon, a schema that did not previously exist either in the steam engine, or in the gas discharge tube; in the steam engine both the boiler producing gas under pressure and the source of heat were external to the cylinder; in the gas engine, it is the cylinder itself that, as a combustion chamber, becomes both boiler and firebox: combustion takes place inside the cylinder, it is an internal combustion; in the gas discharge tube, electrodes were indifferent, as conductance remained symmetrical; the discovery of the thermoelectronic effect allows for the construction of a tube that is analogous to the gas discharge tube in which the electrons are polarized, and which in turn renders conductance asymmetrical. The beginning of a lineage of technical objects is marked by this synthetic act of invention constitutive of a *technical essence.*

A technical essence can be recognized by the fact that it remains stable across the evolving lineage, and not only stable, but also productive of structures and functions through internal development and progressive saturation; this is how the technical essence of the internal combustion engine was able to become that of the diesel engine, through additional concretization of its functioning: in the previous carburized engine, the heating of the carburized mix in the cylinder at the moment of compression is nonessential or even detrimental, since it risks producing a detonation instead of producing a deflagration (combustion with a progressive explosive wave), which limits the admissible compression ratio for a given type of fuel; this heating through compression, conversely, becomes essential and positive in the diesel engine, since this is what produces the beginning of deflagration; this positive aspect of the role of compression is obtained through a greater precision of the determination of the moment in which carburetion must intervene in the cycle: in the previous carburized engine, carburetion could take place at any indeterminate moment before the introduction of the carburized mix into the cylinder: in the diesel engine, carburetion must take place after the introduction and compression of pure air, free of fuel vapors, at the moment in which the piston passes the top dead-center, because this introduction provokes the beginning of deflagration (the beginning of the power stroke in the cycle) and cannot provoke it unless it takes place at the moment in which, at the end of compression, the air attains its highest temperature; the introduction of fuel into the air (carburetion) is thus far more charged with functional significance in the diesel engine than in the gasoline engine; it is being integrated into a more saturated, more rigorous system, which allows the manufacturer less freedom and the user less tolerance. The triode is also a system that is more saturated than the diode; in the diode, asymmetrical 54 conductance is limited only by thermoelectric emission: when the cathode-anode voltage is increased, internal current increases more and more for a given temperature of the cathode, but reaches a ceiling (saturation current), which corresponds to the fact that all electrons emitted by the cathode are captured by the anode. One can thus regulate the current traversing a diode only by varying anodic voltage; on the contrary, the triode is a system in which one can vary the current traversing the anode-cathode space in a continuous manner without varying the anode-cathode voltage; the primitive property (the variation of current in direct relation to the anode-cathode voltage) subsists, but is accompanied by a second possibility of variation, which is determined by the command grid voltage; the function of variation that primitively adhered to the anode's voltage, becomes an individualized, free and defined property, which adds an element to the system and consequently saturates it, since the regime of causalities now comprises an additional component; this saturation of the system through the segregation of functions becomes intensified throughout the course of the evolution of a technical object; in the pentode, the current that traverses the cathode-anode space becomes independent of the anode's voltage for values of the anode's voltage between a very low minimum and a high maximum, defined by the possibility of thermal dissipation; this aspect is stable enough to allow utilization of a pentode as a load resistor in relaxation oscillators* that have to produce linear saw-tooth waves for voltages with horizontal deviation in cathode-ray oscillographs; in this case, the screen voltage, command grid voltage, and the third (suppressor) grid remain fixed. Conversely, in the triode, for a given voltage in the command grid, the anodic current varies according to the anode's voltage: in this sense, the triode

55   is still comparable to a diode, whereas this is no longer the case for the pentode in a dynamic state; this difference results from the fact that in the triode, the anode still plays an ambivalent role as both an electrode that captures electrons (a dynamic role) and as an electrode that creates an electric field (a static role); conversely, in the tetrode or pentode, the maintenance of the electric field, which regulates the flow of electrons, is taken care of by the command grid, which plays the role of an electrostatic anode; the anode plate simply retains the role of capturing electrons; for this reason, the gain of the pentode can be much greater than that of the triode, because the function of maintaining the field of electrostatic acceleration is ensured without variation or weakening (since the screen has a fixed potential), even when the anodic voltage lessens as the current increases, because of insertion of a load resistor in the anodic circuit. One could say that both the tetrode and the pentode eliminate the antagonism that exists in the triode between the function of acceleration of electrons by the anode and the function of capturing the electric charges transported by the electrons, which are accelerated by the same anode, a function which entails a drop in anodic potential when a load resistor is introduced and diminishes the acceleration of electrons. From this point of view, the screen grid must be considered an electrostatic anode with fixed voltage.

One can therefore see that the tetrode and pentode are both the result of the saturation and synergetic concretization of the triode's primitive schema. The screen-grid concentrates within itself all of the functions relating to the electrostatic field, which correspond to the preservation of a fixed potential; the command grid and anode preserve only the functions relating to a variable potential, which they can thus fulfill to a greater extent (during functioning the anode of a pentode deployed as a voltage amplifier can be brought to potentials varying between 30 and 300 volts in a dynamical state); the command grid captures fewer electrons

56   than a triode, which enables one to treat its input impedance as very high: the command grid increasingly becomes a pure command grid no longer exposed to the continuous current created by the captivating of electrons; it is, more rigorously speaking, an electrostatic structure. One can thus consider the pentode and tetrode as direct descendants of the triode, since they realize the development of its internal technical schema by reducing incompatibilities via the redistribution of functions into synergetic subsystems. It is the latency and stability of the concrete schema of organizational invention throughout its successive developments that ground the unity and distinctiveness of a technological lineage.

Concretization gives the technical object an intermediate place between the natural object and scientific representation. The abstract technical object, in other words the primitive technical object, is far from constituting a natural system; it is the translation into matter of a set of notions and scientific principles that are deeply separate from one another, which are attached only through their consequences and converge for the purpose of the production of a desired effect. This primitive technical object is not a natural, physical system, it is the physical translation of an intellectual system. For this reason, it is an application or a bundle of applications; it comes after knowledge, and cannot teach anything; it cannot be examined inductively like a natural object, precisely because it is artificial.

On the contrary, the concrete technical object, which is to say the evolved technical object, comes closer to the mode of existence of natural objects, tending toward internal coherence, toward a closure of the system of causes and effects that 57 exert themselves in a circular fashion within its bounds, and it moreover incorporates a part of the natural world that intervenes as a condition of functioning, and is thus part of the system of causes and effects. As it evolves, this object loses its artificial character: the essential artificiality of an object resides in the fact that man must intervene to maintain the existence of this object by protecting it against the natural world, giving it a status of existence that stands apart. Artificiality is not a characteristic denoting the fabricated origin of the object in opposition to spontaneous production in nature: artificiality is that which is internal to man's artificializing action, whether this action intervenes on a natural object or on an entirely fabricated one; a flower, grown in a greenhouse, which yields only petals (a double flower) without being able to engender fruit, is the flower of an artificialized plant: man diverted the functions of this plant from their coherent fulfillment, to such an extent that it can no longer reproduce except through procedures such as grafting, requiring human intervention. Rendering a natural object artificial leads to the opposite results to that of technical concretization: the artificialized plant can only exist in a laboratory for plants, the greenhouse, with its complex system of thermal and hydraulic regulations. Its system of primitively coherent biological functions has opened up into functions that are independent of one another, and only become attached to one another through the gardener's care; its flowering has become a pure flowering, detached, anomic; the plant flowers until it is exhausted, without producing seeds. It loses its initial capacity of resistance against cold, drought, and sun; the regulations of the primitively natural object become the artificial regulations of the greenhouse. Artificialization is a process of abstraction within the artificialized object.

Conversely, technical concretization makes the primitively artificial object increasingly similar to a natural object.[4] This object needed a regulative external milieu in the beginning, the laboratory, workshop, or sometimes the factory; it gradually increases its concretization, it becomes capable of doing without the artificial milieu, because its internal coherence increases, its functional systematicity closes as it organizes itself. The concretized object is comparable to the spontaneously produced object; the object frees itself from the originally associated laboratory and dynamically incorporates the laboratory into itself through the play of its functions; what enables the self-maintenance of the object's conditions of functioning is its relation to other technical and natural objects, and it is this relation that becomes regulative; this object is no longer isolated; it associates itself with other objects, or suffices unto itself, whereas at first it was isolated and heteronomous.

The consequences of this concretization are not only human and economical (allowing decentralization, for example), they are also intellectual: since the mode of existence of the concretized technical object is analogous to that of natural spontaneously produced objects, one can legitimately consider them as one would natural objects; in other words, one can submit them to inductive study. They are no longer mere applications of certain prior scientific principles. By existing, they prove the viability and stability of a certain structure that has the same status as a natural structure, even if it might be schematically different from all natural structures. The study of the functioning of concrete technical objects bears scientific value, since its objects are not deduced from a single principle; they are testimony to a certain mode of functioning and compatibility that exists in fact and has been built before having been planned: this compatibility was not contained in each of the separate scientific principles that served to build the object; it was discovered empirically; one can work backward from the acknowledgement of this compatibility to the separate sciences in order to pose the problem of the correlation of their principles and ground a science of correlations and transformations that would be a general technology or mechanology.

However, for this general technology to make sense, one must avoid the improper identification of the technical object with the natural object and more specifically with the living being. External analogies, or rather resemblances, must be rigorously banned: they have no signification and are only misleading. Dwelling on automata is dangerous because it risks limiting one to the study of external aspects and thereby to improper identifications. The only thing that counts is the exchange of energy and information within the technical object or between the

---

4. Variant: the object frees itself and becomes naturalized. — Ed.

technical object and its milieu; external behaviors as viewed by a spectator are not objects of scientific study. One needn't even found a separate science that would study the mechanisms of regulation and command in automata built to be automata: technology must deal with the universality of technical objects. In this sense, cybernetics is insufficient: it has the immense merit of being the first inductive study of technical objects, and of presenting itself as a study of the intermediate domain between the specialized sciences; but it has specialized its domain of investigation too narrowly, because it started from the study of a certain number of technical objects; it accepted as its point of departure that which technology must reject: a classification of technical objects according to criteria established according to genera and species. Automata are not a *species*; there are only technical objects, which in turn have a functional organization that results in various degrees of automatism.

What risks making the work of cybernetics partially inefficient as an inter-scientific study (which nevertheless is the objective Norbert Wiener attributes to his research) is the initial postulate concerning the identity between living beings and self-regulating technical objects. Yet the only thing we can say is that technical objects tend toward concretization, whereas natural objects, such as living beings, are concrete to begin with. One mustn't confuse the tendency toward concretization with the status of entirely concrete existence. To a certain extent, every technical object has residual aspects of abstraction; one mustn't go right to the limit and speak of technical objects as if they were natural objects. Technical objects must be studied in their evolution in order to discern the process of concretization as a tendency; but one mustn't isolate the last product of technical evolution in order to declare it entirely concrete; it is more concrete than the preceding ones, yet it is still artificial. Instead of considering one class of technical beings, automata, one must follow the lines of concretization throughout a temporal evolution of technical objects; it is only by following this path that the rapprochement between the living being and the technical object makes any true sense, beyond any mythology. In the absence of any end-point thought out and realized by living human beings on Earth, physical causality could not, in the majority of cases, have produced a positive and efficient[5] concretization on its own, even though modulating structures exist in nature (relaxation oscillators, amplifiers) — wherever metastable states exist, and this is perhaps one of the aspects of the origins of life.

60

---

5.   The end of this sentence is a correction intended for the 1958 manuscript. — Ed.

CHAPTER TWO

EVOLUTION OF TECHNICAL REALITY; ELEMENT, INDIVIDUAL, ENSEMBLE

I. – Hypertely and self-conditioning in technical evolution

There are phenomena of hypertely that manifest themselves in the evolution of technical objects, giving each technical object an exaggerated degree of specialization and maladapting it to even a slight change in the conditions of its utilization or fabrication; the schema that constitutes the essence of the technical object can in fact adapt in two ways: it can first of all adapt to the *material and human conditions* of production; each object can utilize the electrical, mechanical, or chemical aspects of the materials that constitute it in the best possible way; it can then adapt to *the task* for which it is made: a tire that can be used well in a cold climate may not be appropriate for a hot climate and vice versa; a plane made for high altitudes may be disadvantaged by the need to function temporarily at low altitudes, in particular for landing and takeoff. The jet engine, which is superior in very high altitudes to a propeller engine precisely because of its principle of propulsion, becomes difficult to use at a very low altitude; the great speed attained by a jet engine becomes a rather paralyzing aspect when it needs to touch ground; the reduction of wing area, which comes with the use of a jet engine, requires landing at a high speed (almost the cruising speed of a propeller aircraft), thus requiring a very long landing strip.

The first planes, which were able to land in the middle of the countryside, were less functionally over-adapted than modern planes. Functional over-adaptation goes so far that it results in certain schemas similar to those which in biology, fluctuate between symbiosis and parasitism: some very fast small planes can take off with ease only if they are carried by a bigger plane that drops them in mid-flight; others use rockets in order to increase upward thrust. The transport glider is itself a hypertelic technical object: it has become little more than an air cargo ship or rather an air-barge without a tow plane, and as such, it is entirely different from a true glider that can, after a light launch, catch the air on its own, using air currents. The autonomous glider is very finely adapted to engineless flight, whereas the transport glider is only one of two asymmetrical halves of a technical totality, the other being the tow plane; the tow plane, on the other hand, is maladapted because it cannot, on its own, take off with a load corresponding to its power.

One could thus say that there are two types of hypertely: one corresponds to a fine-tuned adaptation to well-defined conditions without breaking the technical object up and without a loss of autonomy; the other corresponds to a breaking-up of the technical object, as in the case of the division of a unique primitive being into tower and towed. The first case preserves the autonomy of the object, whereas the second case sacrifices it. A mixed case of hypertely is one that corresponds to an adaptation to the milieu, such that the object necessitates a certain kind of milieu in order to function properly, because it is energetically coupled to its milieu; this case is almost identical to that of the division into tower and towed; for instance, a clock synchronized by a grid loses all capacity for functioning if it is brought from America to France, because of the different frequency (60 Hertz and 50 Hertz); an electrical motor requires a grid or generator; a synchronous single-phase motor is more finely adapted to a determinate milieu than a universal motor; in this milieu, it functions even better, but outside of this milieu it becomes worthless. A synchronous three-phase motor is even more finely adapted to functioning in a certain type of grid than a single-phase motor, but outside of this grid it can no longer be used; as a result of this limitation its functioning becomes even more satisfactory than that of a single-phase motor (a more regular regime of functioning, high efficiency, very little wear and tear, and low losses in the connecting lines).

In some cases, this adaptation to the technical milieu is primordial; the utilization of an alternative three-phase current is thus fully satisfactory in the factory environment and for all kinds of motors regardless of their power requirements. And yet, no one has been able to this day to use alternative three-phase current for the traction of electric trains. A system of transfer is needed to connect and mutually adapt the locomotive motor using direct current to the three-phase high-voltage transportation grid: it is either the sub-station delivering direct voltage to the feeders of the catenary or the transformers and rectifiers on the locomotive that provide the motor with direct current, even though the catenary is subject to alternative voltage. The locomotive motor would in fact have been constrained to lose too great a part of its range of utilization by adapting in terms of its energy and frequency to the energy distribution grid; a synchronous or asynchronous motor provides large quantities of mechanical energy only when it has reached its design 64 speed; while excellent for a stationary machine like a lathe or a drill that starts up with zero load and is only required to overcome an important degree of resistance once the design speed is reached, this type of utilization is not suitable for the motor of a locomotive; the locomotive starts up with a full load, with all the inertia of its train; it is when it is functioning at operating speed (if one can speak of a system's operating speed, strictly speaking, in the case of a locomotive) that it has the least amount of energy to provide; the motor of a locomotive must provide maximum energy in its transition phases, either during acceleration, or deceleration, or for braking using counter-current. Rich in frequent adaptation to the variations of the operating system, such usage is opposed to the reduction of the spectrum of the modes of utilization characteristic of the adaptation to a technical milieu, for which the factory with its polyphase grid and constant frequency is an example. The example of the traction motor allows us to grasp the existence of a twofold relation that the technical object entertains, with its geographical milieu on the one hand, and its technical milieu on the other.

The technical object is situated at the meeting point between two milieus, and it must be integrated to both milieus at once. In any case, as these two milieus are two worlds that do not belong to the same system and are not necessarily completely compatible, the technical object is to a certain extent determined by human choice, attempting to realize the best possible compromise between these two worlds. In a sense, a traction motor, like a factory motor, is what is fed by the energy of high voltage alternative three-phase lines; in another sense, it is what deploys its energy to pull a train, from stop to full speed and once again to a full halt, via degrees of decreasing speed; it is what must pull the train on its ramps, through turns, and on slopes, while maintaining its speed as constant as possible. 65 The traction motor not only transforms electrical energy into mechanical energy; it applies it to a varíed geographical world, which translates technically into the shape of the tracks, the variable resistance of the wind, the resistance of snow that the front of the locomotive pushes out of the way. The traction motor's reaction rebounds on the line that feeds it, creating a reaction that is the translation of this geographical and meteorological structure of the world: the absorbed intensity increases and the voltage in the line decreases when the snow thickens, when the slope rises, when lateral wind pushes the wheels' guards against the tracks and increases friction. *The two worlds act upon each other* via the traction motor. A three-phase factory motor, on the contrary, does not establish a reciprocal relation of causality between the technical world and the geographical world in the same way; its operation takes place almost entirely within the technical world. The uniqueness of this milieu explains why there is no need for an adaptation milieu in the case of the factory motor, whereas the traction motor requires an adaptation milieu such as is constituted by the rectifiers, which are placed in the sub-station or on top of the locomotive; the factory motor requires only the transformer as an adaptation milieu that lowers voltage, which could be done away for high powered engines, and is necessary in the case of average power engines, as a safety requirement aimed at human users, rather than as a true adaptor to the milieu.

Adaptation follows a different curve and has a different sense in this third case; it cannot lead as directly to phenomena of hypertely and maladaptation following hypertely. The necessity of adaptation, not to a milieu defined as exclusive, but to the function of relating two milieus that are both evolving, limits adaptation and gives it more precision in the direction of autonomy and concretization. This is where true technical progress resides. Thus the use of silicon plates, having greater magnetic permeability and lower hysteresis than iron plates, facilitates a reduction of both the weight and volume of traction motors even while it increases their efficiency; such modifications tend toward a function of mediation between the technical and geographical worlds, since a locomotive can now have a lower center of gravity, with its motors often located at the level of the bogies; the inertia of the rotor is reduced, which is desirable for the sake of rapid braking. The use of silicon insulation makes greater heat tolerable without the risk of deteriorating the insulation, which in turn increases the possibilities of over-intensity, increasing the motor torque during start-up and the resistant torque during braking. Such modifications do not restrain, but rather extend the areas whereby traction motors can be used. A motor insulated with silicon can be used without supplementary precaution in a locomotive climbing steep slopes or in a very hot climate; the engine's relational use extends itself; (a small version of) this type of improved motor can be used as a truck speed-reducer; the motor is in fact adapted to the relational modality as such and not only to this precise type of relation that links the grid and the geographical world for the purpose of the train's traction.

An analogous example of concretization is the Guimbal turbine[6]; this turbine is immersed in the penstock and directly coupled to a very small generator contained in a crankcase filled with pressurized oil. The dam wall thus contains an entire electrical factory within the penstock, since the only thing that appears at ground-level is the gatehouse containing the oil reservoir and measuring instruments. The water becomes pluri-functional: it conveys energy by activating the turbine and generator and also transfers heat from the generator; the oil is also remarkably pluri-functional: it lubricates the generator, insulates the windings, and transfers the generated heat from the winding to the crankcase where it is evacuated by the water; lastly, it prevents the seepage of water into the crankcase through the shaft packings, since the pressure of the oil in the box is greater than the pressure of the water outside the box. This over-pressure is itself pluri-functional; under permanent pressure it greases the bearings while at the same time preventing the water from seeping into the bearings if they fail to be watertight. However, it is worth noting that it is as a result of this pluri-functionality that this concretization and relational adaptation became possible. To put the generator into a penstock containing the turbine was unthinkable prior to Guimbal's invention, because even if the problem of water-tightness and insulation were supposedly solved, the generator was too large to be lodged within the conduit; it is the mode of resolution of the problems of water-tightness and electric insulation that allows for the insertion of the generator into the conduit, while also facilitating excellent cooling through the double intermediary of both oil and water. One could go so far as to say that the insertion of the generator into the conduit *renders itself possible* by simultaneously authorizing the energetic cooling by water. Additionally, the great efficiency of cooling allows for a considerable reduction of size while maintaining the same power. The Guimbal generator would be rapidly destroyed by heat if it were used at full load and out in open air, whereas it barely registers any increase in temperature within its double pool of concentric oil and water, both of which pulsate energetically, the oil according to the generator's movement of rotation, and the water according to the turbulence of the turbine. Concretization is here conditioned by an invention that *presupposes the problem to be resolved*; indeed it is due to the new conditions created by concretization that this concretization is possible; the only milieu in relation to which there is non-hypertelic adaptation, is the milieu created by adaptation itself; here the act of adaptation is not merely an

67

68

---

6.   These turbines are of the same type as the bulb units used in recent French tidal power schemes. They are reversible and can pump water during low tide with low energy expense.

act of adaptation in the sense in which this word defines the adaptation to a milieu that is already given prior to the process of adaptation.

Adaptation-concretization is a process that conditions the birth of a milieu rather than being conditioned by an already given milieu; it is conditioned by a milieu that only exists virtually before invention; there is invention because there is a leap that takes place and is justified by means of the relation that it brings about within the milieu that it creates: the condition of possibility of this turbo-generator couple is its realization; it can only be geometrically situated in the conduit if it is physically situated in such a way that it realizes the thermal exchanges that allow for a reduction of its size. One could say that a concretizing invention realizes a techno-geographic milieu (in this case the oil and water in turbulence), which in turn is a condition of possibility of the technical object's functioning. *The technical object is thus its own condition, as a condition of existence of this mixed milieu* which is simultaneously both technical and geographical. This phenomenon of self-conditioning defines the principle according to which the development of technical objects is made possible without a tendency toward hypertely and then mal-adaptation; hypertely occurs when adaptation is relative to a given that exists prior to the process of adaptation; such adaptation effectively seeks conditions that always outpace it, because its reaction does not impact them and in turn doesn't condition them.

69   The evolution of technical objects can only become progress insofar as these tech-nical objects are free in their evolution and not pushed by necessity in the direction of a fatal hypertely. For this to be possible, the evolution of technical objects must be constructive, which is to say that it calls forth the creation of this third tech-no-geographic milieu wherein each modification is self-conditioned. Indeed, this is not about a form of progress conceived as a march in a direction fixed in advance, nor about a humanization of nature; this progress could just as well present itself as a naturalization of man; between man and nature a techno-geographical milieu arises that only becomes possible through man's intelligence: the self-conditioning of a schema as a result of its functioning necessarily requires the use of an inventive function of anticipation, which cannot be found in nature or in already constituted technical objects; thus a vital work [*une œuvre de vie*] is required to take the leap beyond a given reality and its current systematization, toward new forms that only maintain themselves because they exist all together as a constituted system; when a new organ appears in the evolving series, it maintains itself only if it realizes a systematic and pluri-functional convergence. The organ is its own condition. It is in a similar manner that the geographical world and the world of already existing

technical objects enter into a relation in which concretization is organic, and which defines itself through its relational function. Like an arch that is stable only once it is finished, this object that fulfills a function of relation maintains itself and is coherent once it exists and because it exists; it creates its own associated milieu from itself and is really individualized in it.

II. – Technical invention: ground[7] and form   70
in the living and in inventive thought

We can therefore affirm that the individualization of technical beings is the condition of technical progress. This individualization is made possible by the recurrence of causality within a milieu that the technical object creates around itself and that conditions it, just as it is conditioned by it. This simultaneously technical and natural milieu can be called an associated milieu. It is that through which the technical object conditions itself in its functioning. This milieu is not fabricated [*fabriqué*], or at least not fabricated in its totality; it is a certain regime of natural elements surrounding the technical being, linked to a certain regime of elements that constitute the technical being. The associated milieu mediates the relation between technical, fabricated elements and natural elements, at the heart of which the technical being functions. Such is the case of the ensemble constituted by oil and water moving in and around the Guimbal turbine. This ensemble is concretized and individualized by recurrent thermal exchanges that take place within it: the faster the turbine spins, the more there is an increase in the heat generated by the generator through magnetic losses and the Joule effect; but the faster the turbine spins, the greater the increase in the turbulence of the oil around the rotor and that of the water around the crank-case, thereby activating the thermal exchanges between rotor and water. It is this associated milieu that is the condition of existence for the invented technical object. The only technical objects that can be said to have been invented, strictly speaking, are those that require an associated milieu in order to be viable; these cannot in fact be constituted part by part via the phases of successive evolution, because they can exist only as a whole or not at all. Technical objects that essentially put into play a recurrent causality in their relation with the natural

---

7.   The phrase "fond et forme" nearly always means "content and form", and it is important to be aware that Simondon is here employing this typical expression, but changing the content of its meaning, as soon becomes clear in the text; "fond" here and throughout is rather used in the sense, taken from Gestalt theory, of a "ground" or "background" against which a form or figure can emerge – the constant with reference to which a variable can emerge –, without, however, completely losing from view its primary meaning of "content". [TN]

71   world must be invented rather than gradually developed, because these objects are the cause of the condition of their functioning. These objects are viable only if the problem is solved, i.e., only if they exist together with their associated milieu.

This is why we notice such discontinuity in the history of technical objects, with absolute origins. Only a thought that is capable of foresight and creative imagination can accomplish such a reverse conditioning in time: the elements that will materially constitute the technical object and which are separate from each other, without an associated milieu prior to the constitution of the technical object, must be organized in relation to each other according to the circular causality that will exist once the object will have been constituted; thus what is at stake here is a conditioning of the present by the future, by that which is not yet. Such a futural function is only rarely the work of chance; it requires putting into play a capacity to organize the elements according to certain requirements which act as an ensemble, as a directive value, and play the role of symbols representing the future ensemble that does not yet exist. The unity of the future associated milieu, within which the causal relations will be deployed that will enable the functioning of the new technical object, is *represented*, it is *played* or acted out as much as a role can be played in the absence of the true character, by way of the schemes of the creative imagination. The dynamism of thought is the same as that of technical objects; mental schemas react upon each other during invention in the same way the diverse dynamisms of the technical object will react upon each other in their material functioning. The unity of the technical object's associated milieu is analogous to the unity of the living being; during invention, the unity of the living being is the coherence of mental schemes, obtained by the fact that they exist and deploy themselves in the same being; those schemas that are contradictory confront and reduce one another. The reason the living being can invent is because it is an individual being that carries its associated milieu with it; this capacity for conditioning itself lies at the root of the capacity to produce objects that condition themselves. What

72   has escaped the attention of psychologists in their analysis of the inventive imagination aren't the schemas, forms, or operations that stand out as the spontaneously salient and striking elements, but rather the dynamic ground upon which these schemas confront each other and combine, and wherein they participate. Gestalt psychology, while recognizing the function of totalities, attributed force to form; a deeper analysis of the inventive process would no doubt show that what is determinant and plays an energetic role are not forms but that which carries the forms, which is to say their ground; the ground, while perpetually marginal with respect to attention, is what harbors the dynamisms; it is that through which the system of

forms exists; forms do not participate in forms, but in the ground, which is the system of all forms or rather the common reservoir of the formes' tendencies, well before they exist separately and constitute themselves as an explicit system. The relation of participation that links forms to ground is a relation that bestrides the present and diffuses an influence of the future onto the present, of the virtual onto the actual. For the ground is the system of virtualities, of potentials, forces that carve out their path, whereas forms are the system of actuality. Invention is the taking charge of the system of actuality through the system of virtualities, the creation of a unique system on the basis of these two systems. Forms are passive in so far as they represent actuality; they become active when they organize in relation to this ground, thereby bringing prior virtualities into actuality. It is no doubt very difficult to shed light on the modalities according to which a system of forms can participate in a ground of virtualities. We can only say that it happens according to the same mode of causality and conditioning as the one which exists in the relation of each of the technical object's structures which are constituted with the dynamisms of the associated milieu; these structures are inside the associated milieu, 73 they are determined by it and, because of it, they are also determined by the other structures of the technical being; they also partially determine it, but each one for its own sake, while the technical milieu, which is determined separately by each structure, in turn determines them altogether by providing the energetic, thermal and chemical conditions of functioning. There is a recurrence of causality between the associated milieu and the structures, but this recurrence is not symmetrical. The milieu plays the role of information; it is the seat of self-regulations, the vehicle of information or of energy that is already governed by information (for instance, the water that is animated by more or less rapid movement, cooling the crank-box more or less rapidly); while the associated milieu is homeostatic, the structures are animated by a non-recurrent causality; each goes in its own direction. Freud analyzed the influence of ground on forms in psychic life by interpreting this influence as one of hidden forms on explicit forms; hence the notion of repression. Experiments have indeed proven that symbolization exists (experiments where a violently emotional scene is described to a subject in a hypnotic state and who, upon waking, recounts this scene by using symbolic transposition), but this does not prove that the unconscious is populated by forms comparable to explicit forms. A dynamics of tendencies is sufficient for explaining symbolization if one considers as effective the existence of a psychic ground, upon which the explicit forms that are generated by the conscious and wakeful state are deployed and in which they participate. It is the milieu associated with a systematics of forms that institutes relations of recurrent causality between these forms and that which causes recast-ings of the system of forms taken as an ensemble. Alienation is the break between ground and forms in psychic life: the associated milieu no longer regulates the dynamism of forms. To date, the imagination has been poorly analyzed because forms have been invested with the privilege of being active and are thought to have the initiative in psychic life and in physical life. In reality there exists a great kin-ship between life and thought: within a living organism all living matter cooperates in life; it is not only the most apparent, or the clearest structures that have the initiative of life in the body; blood, lymph nodes, and conjunctive tissues partake in life; an individual is not only made up of a collection of organs combined with one another into systems; an individual is also made up of that which is neither organ nor structure of living matter, insofar as it constitutes an associated milieu for the organs; living matter is the ground of the organs; it is what allows them to relate to each other and become an organism; it is what maintains the fundamental thermal and chemical equilibriums upon which the organs deliver brisk, but lim-ited variations; the organs participate in the body. The living matter in question is far from being pure indeterminacy and pure passivity; nor is it blind aspiration: it is the vehicle of informed energy. In the same way, thought comprises clear, sepa-rate structures, such as representations, images, certain memories, and certain perceptions. All these elements, however, participate in a ground that gives them a direction, a homeostatic unity, and which acts as a vehicle for informed energy from one to the other and among all of them. One could say that the ground is the implicit axiomatic; in it new systems of forms are elaborated. Without the ground of thought, there would be no thinking being, but rather an unrelated series of discontinuous representations. This ground is the mental milieu associated with the forms. It is the middle term between life and conscious thought, just as the associated milieu of the technical object is the middle term between natural world and the fabricated structures of the technical object. We can create technical beings because we have within us a play of relations and a matter-form relation that is highly analogous to the one we constitute in the technical object. The relation between thought and life is analogous to the relation between the structured tech-nical object and the natural milieu. The individualized technical object is an object that has been invented, i.e., produced through the play of recurrent causality between life and thought in man. An object that has only been thought or only associated with life is not a technical object, but a utensil or apparatus. It has no internal consistency, because it has no associated milieu instituting a recurrent causality.

III. – Technical individualization

The technical object's principle of individualization through recurrent causality within an associated milieu enables us to think with greater clarity about certain technical ensembles and to know whether to treat them as a technical individual or as a collection of organized individuals. We shall speak of a technical individual whenever the associated milieu exists as a condition of functioning *sine qua non*, whereas it is an ensemble in the contrary case. Take a laboratory, such as a laboratory studying the physiology of sensation. Is an audiometer a technical individual? No, not if one considers it independently of the power supply and headphones or speakers used as electro-acoustic transducers. The audiometer is defined, then, by certain requisite conditions of temperature, voltage, and noise levels, so that its frequencies and intensities may be sufficiently stable for the measurements of thresholds. The room's coefficient of absorption, its resonance at this or that frequency must be taken into account; the room is part of the complete apparatus: for the audiometer to operate properly it requires either a flat and barren landscape, or that its measurements be made in an anechoic chamber, with a suspended anti-microphonic floor and a thick layer of glass-wool on the walls. What then is the audiometer in itself, such as it is sold by a manufacturer or such as one makes oneself? It is an ensemble of technical forms, wherein each has a relative individuality; in general it has two high-frequency oscillators, where one is fixed and the other variable; the lower of the two frequency beats serves to produce audible sound; a fader facilitates the dosage of the intensity of the stimuli. Neither of these oscillators constitutes a technical object in itself, because in order for it to be stable the technical object requires the stabilized voltage of both heater and anode. This stabilization is generally obtained through an electronic system with a recurrent causality that functionally constitutes the associated milieu of the oscillators' technical forms; this associated milieu, however, is not entirely an associated milieu; it is, rather, a system of transfer, a means of adaptation enabling the oscillators not to be conditioned by the natural and technical external milieu; this milieu would become a truly associated milieu only if a random drift in frequency in one of the oscillators were, as a consequence, to entail variation in the supply voltage opposing this frequency drift; there would be an exchange through reciprocal causality between regulated supply and the oscillators; it is the ensemble of technical structures that would thereby be self-stabilized; here, on the contrary, only the supply is self-stabilized and it does not react upon random variations in the frequency of one of the oscillators.

76

There is a big theoretical and practical difference between these two cases; if the supply is indeed simply stabilized without a recurrent causal link with the oscillators, then simultaneous utilizations of this supply can be limited or extended without inconvenience; it is then possible to plug a third oscillator into the same supply without disturbing its functioning, provided the normal limits of output are not exceeded; however, in order to obtain an efficient feedback regulation the exact opposite is required: no more than a single structure can be attached to a single associated milieu; otherwise opposed random variations in two structures that are linked non-synergistically with the same milieu might compensate for one another rather than resulting in a regulatory reaction; the structures attached to the same associated milieu have to function synergistically. For this very reason the audiometer comprises at least two distinct parts that cannot be stabilized by the same associated milieu: the frequency generator on the one hand and the amplifier-fader on the other. An action by one of these ensembles on the other must be avoided, which notably leads to the careful separation of both their power supplies and to electrically and magnetically shielding the partition that separates them, in order to avoid any interaction. The material limit of the audiometer, however, is not a functional limit; the amplifier-fader is normally extended by way of the acoustic transducer and by the room or external ear of the subject, depending on whether or not one employs speakers or headphones for coupling with the subject. Henceforth, one can posit the existence of relative levels of individualization of technical objects. This criterion has an *axiological value*: the coherence of a technical ensemble is at its maximum when this ensemble is constituted by sub-ensembles with the same level of relative individualization. In a laboratory of physiology of sensations there is thus no advantage in grouping the audiometer's two oscillators together with the amplifier-fader; however, it is worth grouping the two oscillators together so that both are affected simultaneously and in equal proportion by any variation in voltage or temperature, so that the variation of the lower frequency of the beat, which will result from the two correlated variations of the frequency of each of the oscillators, should be as low as possible, given that the two basic frequencies will rise or fall at the same time. It would be perfectly contrary to the functional unity of the beat-frequency generator, however, if two separate power supplies were used and if one oscillator were plugged into one phase of the mains and the other into another phase. One would thereby cancel the self-stabilizing effect of compensation between two variations, which gives the *ensemble* of the two oscillators its great stability in low beat frequencies. On the other hand, it becomes useful to plug the oscillators into a mains phase different from the one the amplifier-fader is plugged into, so as to avoid any reaction of variations in the amplifier's anodic consumption of the oscillators' power supply.

The principle of individualization of technical objects in an ensemble is thus one of the sub-ensembles with recurrent causality in an associated milieu; all technical objects having recurrent causality within their associated milieu must be separated from one another and connected in such a way as to maintain the independence of these associated milieus from one another. The sub-ensemble of oscillators and that of the amplifier-fader-transducer must therefore be more than simply independent in terms of power supply, they must also be independent in terms of coupling to each other: amplifier input should be of very high impedance in relation to the oscillators' output, so as to ensure that any reaction of the amplifier on the oscillators is very weak. If one were, for instance, to plug the fader directly into the outlet of the oscillators, then the setting of this fader might react on the frequency of the oscillators. An ensemble of a higher degree comprising all these sub-ensembles is defined by the capacity to freely realize any form of relation, without thereby destroying the autonomy of individualized sub-ensembles. This for instance is the role of a general power switchboard and wiring-board in a laboratory; electrostatic and electromagnetic shields, the use of non-reactive coupling such as what we call the *cathode-follower*, aim at maintaining the independence of these sub-ensembles, while also allowing for the diverse combinations among the functioning of sub-ensembles that are necessary; such is the second degree of the functional role of the ensemble that one can call a laboratory, namely the utilization of the results of functioning without any interaction with the conditions of functioning. 79

We might ask then, on what level is the individuality found: at the level of the sub-ensemble or at that of the ensemble? The answer lies, as always, in the criterion of recurrent causality. At the level of the higher ensemble (such as that of the laboratory) there is indeed no truly associated milieu; if it exists, then it does so only in certain respects, and it is not a general milieu; the presence of oscillators in the room where the experiment in audiometry takes place is often problematic; if the oscillators use transformers with an iron magnetic circuit, then the magnetostriction* of the iron plates creates a vibration, which in turn emits an undesirable sound; an oscillator with resistances and capacities also emits a slight sound due to alternative electrical attractions. For fine-tuned experiments it is necessary to place the devices in another room and to operate them by remote control, or to isolate the subject in an anechoic chamber. The same applies to the magnetic radiation of power transformers, which can be very problematic in electro-encephalographic and electrocardiographic experiments. A superior ensemble such as a laboratory is thus above all constituted by un-coupling devices, in order to avoid creating associated milieus by accident. The ensemble distinguishes itself from technical individuals in the sense that the creation of a unique associated milieu is undesirable; the ensemble is comprised of a certain number of devices in order to counteract this possible creation of a unique associated milieu. It avoids internal concretization of the technical objects it contains, and uses only the results of their functioning, without allowing any interaction with their conditioning.

80 Are there other groupings with a certain individuality below the level of technical individuals? — Yes, but this individuality does not have the same structure as that of technical objects with an associated milieu; it is that of a plurifunctional composition without a positive associated milieu, which is to say without self-regulation. Let us take the case of a hot-cathode lamp. When this lamp is inserted into an assembly, with an automatically polarized cathode resistance, then it is indeed the site of phenomena of self-regulation; for instance, if the heating voltage increases, cathodic emission increases, as a result of which negative polarization increases; the lamp's amplification hardly increases nor does its output or its anodic dissipation augment much; a similar phenomenon enables class A amplifiers* to automatically equalize output levels despite variations in the input level of the amplifier. Such regulatory counter-reactions, however, do not uniquely reside within the lamp alone; they depend on the whole of the assembly, and, in some cases, within particular assemblies, they do not exist at all. A diode whose anode heats up thus becomes a conductor in both directions, and this also increases the intensity of the current going through it; the cathode receiving the electrons from the anode heats up even more and emits more electrons: this destructive process thus manifests a positive circular causality partaking in the whole assembly and not only in the diode.

Infra-individual technical objects can be called technical elements; they distinguish themselves from true individuals in the sense that they do not have an associated milieu; they can integrate into an individual; a hot-cathode lamp is a technical element rather than a complete technical individual; it is comparable to an organ in a living body. It would in this sense be possible to define a general 81 organology, studying technical objects at the level of the element, and which would belong to technology, together with mechanology, which would study complete technical individuals.

IV. – Evolutionary succession and preservation
of technicity. Law of relaxation

The evolution of technical elements can have repercussions for the evolution of technical individuals; composed of elements and an associated milieu, technical individuals depend to a certain extent on the characteristics of the elements they implement. Electric magnetic motors can thus be much smaller today than in Gramme's days, because magnets have been greatly reduced in size. In certain cases, elements are like the crystallization of a preceding technical operation that produced them. In this sense, magnets with oriented grains, which we refer to as magnetically tempered, are obtained through a procedure that consists in maintaining a vigorous magnetic field around the molten mass that, once cooled, will become the magnet; starting with the magnetization of the molten mass at a temperature above the Curie point, its intense polarization is then maintained while the mass cools; when the mass is cold, it constitutes a far more powerful magnet than if it had been magnetized after cooling. Everything happens as if the vigorous magnetic field produced an orientation of the molecules in the molten mass, an orientation that maintains itself after cooling, if the magnetic field is sustained during its cooling and transition into the solid state. Now, the furnace, the melting pot, and the coils creating the magnetic field all together constitute a system that is a technical ensemble; the heat of the furnace shouldn't have an impact on the coils, the induction field creating the heat in the molten mass shouldn't neutralize the continuous field aimed at bringing about the magnetization. This technical ensemble is itself constituted of a certain number of technical individuals that are 82 organized in relation to each other, both in view of the result of their functioning and so as to prevent the conditioning of each particular functioning from being disturbed. In the evolution of technical objects we thus witness a passage of causality proceeding from prior ensembles to subsequent elements; such elements, once introduced into an individual whose characteristics they modify, in turn enable technical causality to rise from the level of the element to that of the individual, then from the level of the individual to the level of the ensemble; here, in a new cycle, technical causality once more descends to the level of the element via the process of fabrication, where it reincarnates itself in new individuals and then in new ensembles. There is thus a lineage of causality that is not rectilinear but serrated, where one and the same reality exists first in the form of an element, then as the characteristic of an individual and finally as the characteristic of an ensemble.

The historical solidarity that exists among technical realities is mediated by the fabrication of elements. For a technical reality to have posterity, it is not enough for it simply to improve in itself: it must also reincarnate itself and participate in this cyclical coming-into-being via a process of relaxation within the different levels of reality. The solidarity that exists among technical beings masks this other much more essential solidarity that requires the temporal dimension of evolution, not identical with biological evolution, however, which in turn is not characterized by these successive changes of levels and which occurs along more continuous lines. If transposed into biological terms, technical evolution would consist in the fact that a species could produce an organ that would be given to an individual, which would thereby become the first term of a specific lineage, which, in turn, would produce a new organ. In the domain of life, an organ is not detachable from the species; in the technical domain, an element is detachable from the whole that produced it, precisely because it is fabricated; and here, we see the difference between the *engendered* and the *produced*. In addition to its spatial dimension, the technical world has a historical dimension. Its current solidarity mustn't mask the solidarity of succession; this latter solidarity is in fact what determines the great epochs of technical life through a law of serrated evolution.

83

Nowhere else does such a rhythm of relaxation find its equivalent; neither the human nor the geographical world can produce such oscillations of relaxation, with successive fits and spurts of new structures. This relaxation time is the technical time properly speaking; it can become dominant with respect to all other aspects of historical time, to the extent that it can even synchronize other rhythms of development and appear to determine the entire historical evolution, when in fact it only synchronizes and brings about its phases. An example of this evolution following a rhythm of relaxation can be found in energy sources since the eighteenth century. A large part of energy employed during the eighteenth century came from waterfalls, displacements of atmospheric air and from animals. These types of prime movers corresponded to artisanal exploitation or rather limited factories distributed along waterways. From these artisanal factories emerged the high efficiency thermodynamic machines at the beginning of the nineteenth century, and the modern locomotive, which is the result of the adaptation of Stephenson's valve gear to the multi-tubular boiler designed by Marc Seguin, lighter and smaller than a French boiler. This valve gear allows for the variation of the relation between admission time and expansion time, as well as for the ability to reverse gears (the reversal of steam), through the intermediary of a neutral position. This artisanal type of mechanical invention, which grants to the traction engine the capacity to

apply itself with ample variations of the engine torque to highly varied profiles, 84
at the cost of a loss of efficiency only in high power regimes (where the time of admission is almost equal to the totality of the expansion stroke), makes thermal energy easily adaptable to traction on rails. Stephenson's valve gear and tubular boiler, which are elements that emerge from the artisanal ensemble of the eighteenth century, enter into the new individuals of the nineteenth century, especially through the form of the locomotive. High tonnage transportation, which had now become possible throughout all regions, and which was no longer constrained to the contour lines and meandering of navigable tracks, led to the industrial concentration of the nineteenth century, which not only incorporates individuals whose functioning principle is based on thermodynamics, but is essentially thermodynamic in its structures; it is therefore around coal sources of thermal energy and close to the sites of the greatest deployment of thermal energy (coal mines and iron works) that the great nineteenth-century industrial ensembles, at the peak of their reign, are concentrated. We have gone from the thermodynamic element to the thermodynamic individual and from the thermodynamic individual to the thermodynamic ensemble.

The principal aspects of electrotechnics will in turn emerge as elements produced by these thermodynamic ensembles. Before acquiring their autonomy, the applications of electrical energy emerge as highly flexible means for the transmission of energy from one place to another by way of energy transport cables. Metals with high magnetic permeability are elements that are produced by way of the application of thermodynamics to metallurgy. Copper cables and high resistance porcelains for insulators emerge from steam-powered wire mills and coal furnaces. The metallic frameworks of pylons as well as the cement for dams are both born out of great thermodynamic concentrations and enter as elements into the new technical individuals that the turbines and alternators are. A new wave and a new 85
constitution of beings become emphasized and concretized. In the production of electrical energy Gramme's machine gives way to the polyphase alternator; the direct current of early energy transport gives way to alternative currents with constant frequency, adapted to being produced by thermal turbine and consequently also being produced by hydraulic turbine. These electrotechnic individuals have integrated themselves into the ensembles of the production, distribution and utilization of electric energy, whose structure differs vastly from that of thermodynamic concentrations. The role played by the railway in this thermodynamic concentration is now replaced by the role played by high voltage transmission lines in the ensemble of industrial electricity.

At the moment in which electrical technics reaches its full development, it produces new schemes in the form of elements that initiate a new phase: first, there is particle acceleration, initially realized through electric fields and then through continuous electric fields and alternative magnetic fields, which leads to the construction of technical individuals, having enabled the discovery of the possibility of exploiting nuclear energy; subsequently, and quite remarkably, there is the possibility, afforded by electrical metallurgy, of extracting metals like silicon, which allows the transformation of the radiant energy of light into electrical current, with an efficiency that already reaches a level relevant for limited applications (6%), and which is not much lower than that of the first steam engines. A pure silicon photo-cell, produced by large industrial electrotechnical ensembles, is the element that hasn't yet been incorporated into a technical individual; it is still only an object of curiosity situated at the extreme end of the technical possibilities of the electro-metallurgic industry, but it is possible that it may become the point of departure for a new phase of development, analogous to the one we have experienced with the development of the production and utilization of industrial energy, which itself is not yet complete.

Every phase of relaxation is capable of synchronizing minor aspects or those of almost equal importance; the development of thermodynamics thus went hand in hand with railway transportation, not simply with the transportation of coal but of passengers; on the contrary, the development of electrotechnics went hand in hand with the development of automotive transport; the automobile, albeit thermodynamic in principle, uses electric energy as an essential auxiliary, in particular for ignition. Industrial decentralization, facilitated by long distance electrical energy transfer, needs the automobile as a corresponding means for the transportation of people to locations that are distant from each other and at different altitudes; corresponding to the road rather than railway. The automobile and the high voltage line are parallel technical structures that are synchronized, but not identical: electrical energy cannot currently be applied to automobile traction.

Furthermore, there is no intrinsic relation between nuclear energy and energy derived from the photoelectric effect; these two forms are nevertheless parallel, and their developments are susceptible to mutual synchronization[8]; nuclear energy will thus probably remain, for a long time to come, inapplicable to direct forms of restricted utilization, such as those consuming a few dozen watts; photoelectric energy, on the contrary, is a highly decentralizable energy; it is essentially decentralized in its production, whereas nuclear energy is essentially centralized. The

---

8.   And to conjugate each other: a solar cell can be irradiated by a radioactive source.

relation that existed between electrical energy and energy derived from the combustion of gasoline exists once more between nuclear energy and photoelectric energy, perhaps with a more accentuated difference.

V. – Technicity and evolution of technics:
technicity as instrument of technical evolution

The different aspects of the technical being's individualization constitute the center of an evolution, which proceeds via successive stages, but which is not dialectical in the proper sense of the term, because, in regard to it, negativity does not play the role of an engine of progress. In the technical world negativity is a lack of individuation, an incomplete junction of the natural world and the technical world; this negativity is not the engine of progress; it is rather the engine of transformation, it incites man to seek new, more satisfactory solutions than those he possesses. This desire for change, however, does not happen directly within the technical being; it happens within man as inventor and user; this change moreover must not be confused with progress; a change that is too abrupt is contrary to technical progress, because it prevents the transmission, in the form of technical elements, of what an age has acquired to the one that follows.

For progress to exist, each age must be able to pass on to the next age the fruit borne of its technical effort; what is capable of being passed on from one age to another are neither technical ensembles, nor even individuals, but the elements that these individuals, grouped as ensembles, were able to produce; thanks to their capacity of internal inter-commutation, technical ensembles in fact have the possibility of going beyond themselves by producing elements that differ from their own. Technical beings are different from living beings in many respects, but they differ essentially in the following respects: a living being engenders beings that are similar to itself, or that can become so after a certain number of successive reorganizations occurring spontaneously if the required conditions obtain; a technical being, on the contrary, does not have this capacity; it cannot spontaneously produce other technical beings similar to itself, despite the efforts of cyberneticists who have tended to force technical beings to copy the living by constructing beings that are similar to them: this possibility is currently a mere supposition and is without foundation; but the technical being has greater freedom than the living, afforded to it by its infinitely lesser degree of perfection; in these conditions the technical being can produce elements that harbor the degree of perfection at

88

which the technical ensemble had arrived at, and which, in turn, can be reunited to enable the construction of new technical beings in the form of individuals; there is thus no engendering here, no procession or direct production, but only indirect production through the constitution of elements that contain a certain degree of technical perfection.

This affirmation requires that we specify what the process of technical perfection is. Empirically and externally one can say that technical perfection is a practical quality, or at the very least the material and structural basis of certain practical qualities; in this way a good tool is not simply one that is well put together and well crafted. In practical terms an adze can be in poor condition, blunt, and yet not be a bad tool; an adze is a good tool if, on the one hand, it has a curve suited for a straight, well aimed strike at the wood, and, on the other hand, if it can be sharpened and keep its sharpness even when employed to work on hard wood. This latter quality in turn results from the technical ensemble employed to produce the tool.

89 It is because it is a fabricated element that the adze can be made of a metal whose composition varies at different points; this tool is not only a hunk of metal shaped into a certain form; it has been forged, which is to say that the molecular chains of the metal have a certain orientation that varies in certain places, like a piece of wood whose fibers are oriented in the direction that offers the greatest solidity and elasticity, especially in the intermediary parts between the cutting edge and the flat thick part which goes from the socket to the cutting edge; this region close to the cutting edge deforms itself elastically during work, because it acts as both wedge and lever on the piece of wood being chipped off. And finally, the cutting edge has a higher steel content than the other parts; its steel needs to be hard, but in a proper delimited way, for too much of hard steel in the cutting edge would make the tool brittle and the edge would shatter into splinters. It is as if, in its totality, the tool was made of a plurality of functionally different zones, welded together. The tool is made not only of form and matter; it is made of elaborate technical elements according to a certain schema of functioning and assembled into a stable structure through the operation of fabrication. The tool unites within itself the results of the functioning of a technical ensemble. In order to make a good adze a technical ensemble of a foundry, forge, and quench hardening is required.

The technicity of the object is thus more than a quality of its use; it is that which, within it, adds itself to a first determination given by the relation between form and matter; it acts as an intermediary between form and matter, here for instance as the progressive heterogeneity of the quench hardening in different points. Technicity is the degree of the object's concretization. During the period of wood foundries, it

was this concretization that gave Toledo's blades their value and prestige, and more
recently, led to the quality of Saint-Étienne's steel. These types of steel express the   90
result of the functioning of a technical ensemble comprising in equal measure the
qualities of coal used, as well as the temperature and chemical composition of the
soft water of the Furan river, or the species of green wood used to stir and refine the
molten metal prior to casting. In certain cases, technicity becomes predominant
with respect to the abstract aspects of the relation between matter and form. A coil
spring is thus a very simple thing in form and matter, yet the fabrication of springs
requires a high degree of perfection in the technical ensemble that produces them.
The quality of individuals, such as an engine or an amplifier, often depends much
more on the technicity of simple elements (valve springs, for instance, or a modulation transformer) than on the ingenuity of their assembly. Technical assemblies,
however, that are capable of producing certain simple elements, such as a spring or
a transformer, are sometimes extremely vast and complex, and almost coextensive
with all the ramifications of several global industries. It would not be an exagger-
ation to say that the quality of a simple needle expresses the degree of perfection
of a nation's industry. This explains why there are judgments that are legitimate
enough in both practical and technical terms, such as when a needle is specifically
called an "English needle." Such judgments have a practical sense, because technical ensembles express themselves in the simplest elements they produce. This mode
of thought of course exists for other reasons besides those that legitimate it, and
particularly because it is easier to qualify a technical object by its origin than to
judge its intrinsic value; what we have here is a phenomenon of opinion; but even
if this phenomenon gives rise to numerous exaggerations or intentional exploitation, it is not without foundation.

Technicity can be considered a positive aspect of the element, analogous to
the self-regulation exerted by the associated milieu in the technical individual.
Concretization is technicity at the level of the element; it is the reason why the ele-   91
ment is really an element produced by an ensemble, rather than being an ensemble
itself or an individual; this characteristic makes it detachable from the ensemble
and frees it so that new individuals may be constituted. There is of course no
peremptory reason why one would attribute technicity only to the element; the
associated milieu is the depositary of technicity at the level of the individual, just
as extension is the depositary of inter-commutativity at the level of the ensemble;
it is nevertheless good to reserve the term technicity for this quality of the element,
which expresses and preserves what has been acquired via a technical ensemble so as
to be transported into a new period. What the element transports is a concretized

technical reality, whereas the individual and the ensemble contain this technical reality without being able to transport and transmit it; elements have a transductive property that makes them the true bearers of technicity, just as seeds transport the properties of a species and go on to make new individuals. It is thus within elements that technicity exists in the purest way, in a free state as it were, whereas in the individual or the ensemble, technicity only exists in a state of combination.

However, this technicity, borne by the elements, contains no negativity, and no negative conditioning intervenes in the moment of production of elements by the ensembles or of individuals by invention, which reunites the elements in order to form individuals. Invention, which is a creation of the individual, presupposes in the inventor the intuitive knowledge of the element's technicity; invention occurs at this intermediate level between the concrete and the abstract, which is the level of schemas, and presupposes the pre-existence and coherence of representations that cover the object's technicity with symbols belonging to an imaginative sys-92 tematic and an imaginative dynamic. The imagination is not simply the faculty of inventing or eliciting representations outside sensation; it is also the capacity of the prediction of qualities that are not practical in certain objects, that are neither directly sensorial nor entirely geometric, that relate neither to pure matter nor to pure form, but are at this intermediate level of schemas.

We can consider the technical imagination as being defined by a particular sensitivity to the technicity of elements; it is this sensitivity to technicity, that enables the discovery of possible assemblages; the inventor does not proceed *ex nihilo*, starting from matter that he gives form to, but from elements that are already technical, with respect to which an individual being is discovered as that which is susceptible to incorporating them. The compatibility of elements in a technical individual presupposes the associated milieu: the technical individual must therefore be imagined, which is to say presupposed as already being constructed in the form of an ensemble of ordered technical schemas; the individual is a stable system of the technicities of elements organized as an ensemble. What is organized are these technicities, as well as the elements as basis of these technicities, not the elements themselves taken in their materiality. An engine is an assemblage of springs, axes, and volumetric systems, each defined by their characteristics and their technicity, not by their materiality; thus, a relative indeterminacy can subsist in the localization of this or that element in relation to all the others. The location of certain elements is chosen more by virtue of extrinsic considerations than by intrinsic considerations of the technical object in relation to the diverse processes of its functioning. The intrinsic determinations, based on the technicity of each element, are those which constitute the associated milieu. This associated milieu, in turn, is the concretization of the technicities contributed by all the elements, in their mutual reactions. Technicities can be thought of as stable behaviors, express- 93 ing the characteristics of elements, rather than as simple qualities: they are powers, in the fullest sense of the term, which is to say capacities for producing or undergoing an effect in a determinate manner.

The higher the technicity of an element, the more the margin of indeterminacy of this power diminishes. This is what we want to express when we say that the elementary technical object concretizes itself as its technicity increases. One could thus name this power *a capacity*, if one intends to characterize it in relation to a determinate deployment. Generally, the higher the technicity of an element, the wider the conditions of deployment of this element are, as a result of the high level of stability of this element. The technicity of a spring increases when it is capable of bearing higher temperatures without losing its elasticity, when it preserves its coefficient of elasticity without significant modification within larger thermal and mechanical limits: it technically remains a spring but within a much larger framework, and is suited to a less restricted incorporation into this or that technical individual. An electrolytic condenser* has a lower degree of technicity than a dry dielectric condenser, such as paper or mica. An electrolytic condenser in fact has a capacitance that varies as a function of the voltage to which it is submitted; the thermal limits of its utilization are more restricted. At the same time it varies when submitted to constant voltage, because the electrolytes as well as the electrodes become chemically altered during the course of their functioning. Dry dielectric condensers, on the contrary, are more stable. Nevertheless, the technical quality once again increases with the independence of its characteristics from the conditions of utilization; a mica condenser is better than a paper condenser, and the vacuum condenser is the best of all, since it is not even subject to the condition that the voltage be limited lest the insulation risk perforation; an intermediary 94 degree, the ceramic silver-plated condenser for instance, which hardly varies with temperature, and the air condenser, both provide a high degree of technicity. It must be noted that in this sense there is not necessarily a correlation between the commercial price of a technical object and its elementary technical quality. Very often, considerations of price do not intervene in absolute terms, but via another requirement, like that of space; an electrolytic condenser is thus preferred to a dry dielectric condenser where its high capacity would require too much space to house the condenser; similarly, an air condenser is often cumbersome with respect to a vacuum condenser of the same capacitance; although it is much cheaper, and is equally safe to deploy in a dry atmosphere. Therefore, in many cases, economic considerations do not intervene directly, but through the repercussions that the degree of concretization of the technical object has on its deployment in an individual ensemble. It is the general formula of the individual that is subjected to economic repercussions, not that of the element as element. The liaison between the technical and the economic domains occurs at the level of the individual or the ensemble, but very rarely at the level of the element; in this sense, one could say that technical value is largely independent of economic value and that it can be appreciated according to independent criteria.

This transmission of technicity by its elements is what grounds the possibility of technical progress, above and beyond the apparent discontinuity of forms, domains, the types of deployed energy, and sometimes even beyond the schemas of functioning. Each stage of development is the inheritor of previous ages, and its progress is all the more certain as each stage tends increasingly and more perfectly toward a state of sole beneficiary.

95

The technical object is not directly a historical object: it is subject to the course of time only as a vehicle of technicity, according to a transductive role that it plays with respect to a prior age. Neither the technical ensembles nor technical individuals remain; only elements have the power to transmit technicity from one age to another, in the form of an effectuated, accomplished, materialized result. For this reason it is legitimate to analyze the technical object as consisting of technical individuals; but it is necessary to stress that at certain moments in its evolution the technical element makes sense in itself, and is thus a depositary of technicity. In light of this, one can establish the foundation of the analysis of the technics of a human group through the analysis of elements produced by its individuals and its ensembles: often these elements alone have the power to survive the downfall of a civilization, and remain valid witnesses of a state of technical development. In this sense, the method of ethnologists is perfectly valid; but one could prolong its application by equally analyzing the elements produced by industrial techniques.

Indeed, there is no fundamental difference between a people who have no industry and those who have a well-developed industry. Even in a people without any industrial development, technical individuals and technical ensembles exist; nevertheless, rather than being stabilized by institutions that fix and perpetuate them by installing them, these individuals and ensembles are temporary or even occasional; what is preserved from one technical operation to another are merely the elements, which is to say tools or certain fabricated objects. To build a boat is an operation that requires a truly technical ensemble: a fairly flat ground, yet close to the water, sheltered yet luminous, with supports and wedges to keep the ship standing while it is being built. The shipyard, like the technical ensemble, can be temporary: it is   96 no less a shipyard constituting an ensemble. Even today similar temporary technical ensembles still exist, sometimes even highly developed and complex ones, such as the construction sites of buildings; others are provisional while being durable, like mining facilities or the drilling rigs for oil exploration.

Not all technical ensembles necessarily take on the stable form of the factory or the workshop. On the other hand it seems that non-industrial civilizations differentiate themselves from ours mostly by the absence of technical individuals. This is true only if what is meant is that technical individuals do not exist materially in a stable and permanent way; the function of technical individualization, however, is assumed by human individuals; the process of learning, through which man forms habits, gestures, and schemas of action that enable him to use the highly varied tools that the totality of an operation requires, pushes this man to individualize himself technically; it is he who becomes the associated milieu of these diverse tools; when he masters all of his tools, when he recognizes the moment he must change tools in order to continue working or to use two or three tools at a time, he ensures an internal distribution and self-regulation of the task[9] through his body. In some cases, the integration in the ensemble of technical individuals happens via the intermediary of an association of human individuals working in twos, in threes,   97 or in larger groups; when these groupings do not introduce functional differentiation, then their only direct purpose is to increase the available energy or speed of the work, but when differentiation is called for, they clearly demonstrate the genesis of an ensemble on the basis of men employed as technical individuals rather than as human individuals: this was the case with bow drilling, as described by the authors of classical antiquity; it is still the case with the felling of certain trees; it was also commonly the case, not so long ago, with the use of a two-man cross-cut saw to make planks or rafters; two men working together in an alternating rhythm. This explains why, in some cases, human individuality can be used functionally as the basis of technical individuality; the existence of separate technical individualities is a rather recent development and even appears, in some respects, like an imitation of man by the machine, where the machine remains the more general form of a technical individual. Yet, in reality machines are very dissimilar to man,

---

9.   This is where a certain nobility of artisanal work comes from: man is the bearer of technicity, and work is the only mode of expression of this technicity. The imperative to work translates this requirement of expression; to refuse to work when one possesses a technical knowledge that can only be expressed through work, because it cannot be formulated in intellectual terms, would be to hide one's light under a bushel. The requirement of expression, on the contrary, is no longer linked to work when technicity has become immanent to a knowledge that can be formulated abstractly and outside of all concrete actualization.

and even when they function in a way that produces comparable results, it is very rare that they use procedures identical to the work of an individual man. The analogy is in fact most often very external. Yet if man often feels frustration before the machine, it is because the machine functionally replaces him as an individual: the machine replaces man as tool bearer. In the technical ensembles of industrial civilizations, jobs where several men must work in narrow synchronization are becoming rarer than in a past characterized by the artisanal level. Conversely, at the artisanal level, it is very frequent that certain works require the grouping of human individuals with complementary functions: to shoe a horse, one man is needed to hold the hoof and another to hold the shoe up and nail it on. In order to build, the mason has his assistant, the hod-carrier. In order to thresh with the flail, one needs a proper perception of the rhythmical structures that synchronize the alternate movements of the team's members. Yet one cannot affirm that what has been replaced by machines are only the helpers; it is the very basis of technical individualization that has changed: this basis was a human individual; it is now the machine; tools are borne by the machine, and one could even define the machine as that which bears and directs tools. Man directs and adjusts or regulates the machine, the tool bearer; he realizes groupings of machines, but does not himself bear tools; indeed, the machine accomplishes the core work, the work of the blacksmith and not that of the helper; man, disengaged from this function of the technical individual, which is the very essence of the artisanal function, can now become either organizer of the ensemble of technical individuals, or helper of technical individuals: he greases, cleans, removes detritus and burrs, in other words, in some respects he plays an auxiliary role; he provides the machine with elements, changing the belt, sharpening the drill or the lathe cutting tool. There is thus, in this sense, a role above that of the technical individuality, and one below it: servant and regulator, he supervises the machine, the technical individual, by looking after the relation of the machine with the elements and the ensemble; he is the organizer of relations between technical levels, rather than being himself, like the craftsman, one of the technical levels. A technician therefore adheres less to his professional specialization than does a craftsman.

This nevertheless does not mean that man cannot be a technical individual in any shape or form and work in conjunction with the machine; this machine-man relation is realized when man applies his action to the natural world through the machine; the machine is then a vehicle for action and information, in a relation with three terms: man, machine, and world, the machine being that which is between man and world. In this case, man preserves some traits of technicity defined in particular by the necessity of apprenticeship. The machine thus essentially serves the purpose of a relay, an amplifier of movements, but it is still man who preserves within himself the center of this complex technical individual that is the reality constituted by man and machine. One could say, in this case, that man is the bearer of the machine, while the machine remains the tool bearer; this relation is thus partially comparable to that of the machine-tool, if what is understood as machine-tool is that which has no self-regulation. It is man who is at the center of the associated milieu in this relation; the machine-tool is that which has no internal autonomous regulation, and which requires man in order to make it work. Man intervenes here as a living being; he uses his own sense of self-regulation in order to operate that of the machine, even without the need for it to be consciously formulated: in order to restart an overheating car engine a man will allow it to "rest," and in order to start it back up, instead of beginning by revving it up too much, he will progressively get it started from a cooler state. Such behaviors, which are technically justified, have their correlation in vital regulations, and are lived by the driver more than simply being thought by him. They apply all the better to the technical object as the latter approaches the status of a concrete being, encompassing homeostatic regulations within its functioning. For the technical object that has become concrete, there is indeed a regime in which the processes of self-destruction are reduced to a minimum, because of the greatest possible degree of perfection in homeostatic regulation. This is the case for the diesel engine, which requires a definite operating temperature and a regime of rotation confined within a narrow margin between minimum and maximum, while the gasoline engine is more flexible, because it is less concrete. Similarly, an electronic tube cannot function with a cathode at any temperature whatsoever or with an indeterminate anodic voltage; for power 100 tubes in particular, too low a cathode temperature causes the electric field to snatch electron emitting oxide particles; hence the need for a gradual start up, beginning with a warm up of the cathodes without anodic voltage, followed by the powering of the anodes. If the circuits of polarization are automatic (fed by the cathodic current), then they must be progressively powered through the gradual feeding of the anodes; omitting this precaution leads to a short instant in which cathodic current already occurs before polarization reaches its normal level (polarization produced by this current and proportional to it tends to limit it): the cathodic output, which is not yet limited by this negative reaction, would exceed the admissible maximum.

Speaking in very general terms, the precautions man takes for the preservation of the technical object have as their goal the maintenance of its functioning or its adjustment to conditions that prevent it from being self-destructive, which is to say to conditions where it exercises a negative stabilizing reaction upon itself; beyond certain limits, reactions become positive, and consequently destructive; this is the case with an engine which, when over-heated, initiates galling and which, in heating up even more because of the heat from the galling, subsequently deteriorates irreversibly; an electronic tube whose anode becomes red hot, similarly loses its asymmetrical conductivity, in particular within its rectifier function: it then enters a phase of positive reaction. Allowing it the proper time to cool enables it to return to its normal functioning.

Thus, man can intervene as a substitute for the technical individual, and connect elements with ensembles, in an age when the construction of technical individuals is not possible.

101   What one must take into account when thinking about the consequences of technical development in relation to the evolution of human societies, is first and foremost the process of the individualization of technical objects; human individuality is increasingly disengaged from the technical function through the construction of technical individuals; for man, the functions that remain are both below and above the role of tool bearer, oriented both toward the relation with elements and toward the relation with ensembles. However, as it was precisely the individuality of man that was once employed in technical work, and which had to technicize itself because the machine couldn't, it became customary to give each human individual just one function in regard to work; this functional monism was perfectly useful and necessary when man became a technical individual. But it now creates unease, because man, who still seeks to be a technical individual, no longer has a stable place alongside the machine: he becomes the servant of the machine or the organizer of the technical ensemble; yet, in order for the human function to make sense, it is necessary for every man employed with a technical task to surround the machine both from above and from below, to have an understanding of it in some way, and to look after its elements as well as its integration into the functional ensemble. For it is a mistake to establish a hierarchical distinction between the care given to elements and the care given to ensembles. Technicity is not a reality that can be hierarchized; it exists as a whole inside its elements and propagates transductively throughout the technical individual and ensembles: through the individuals, ensembles are made of elements, and from them elements issue forth. The apparent pre-eminence of ensembles comes from the fact that ensembles are currently given the same prerogatives as those of people playing the role of the boss. Yet ensembles are not in fact individuals; the devaluation of elements equally results from the fact that the use of elements has hitherto been

proper to helpers and that these elements were not very elaborate. The unease of 102
this situation relating to man and machine thus comes from the fact that one of the
technical roles, that of the individual, has thus far and until this day been played
by men; no longer a technical being, man is, henceforth obliged to learn a new
function and to find a place within the technical ensemble that no longer corresponds to the technical individual; the first movement consists in occupying two
functions that are not individual, that of elements and that of the direction of the
ensemble; yet in these two functions man finds himself to be in conflict with the
memory of himself: man has for so long played the role of the technical individual
that the machine, once it has become a technical individual, still appears like a man
occupying the place of another man, when it is, on the contrary, man who in fact
provisionally replaced the machine before truly technical individuals could emerge.
In all judgments made about the machine, there is an implicit humanization of the
machine whose deepest source lies in this change of role; man had learned to be a
technical being to the point of believing that the technical being, once it becomes
a concrete being, begins illegitimately to usurp the role of man. Ideas of servitude
and liberation are far too strongly related to the old status of man as a technical
object for them to correspond to the true problem of the relation between man
and machine. It is necessary, that the technical object be known in itself, in order
for the relation between man and machine to become stable and valid: hence the
necessity for a culture of technics.

ILLUSTRATIONS

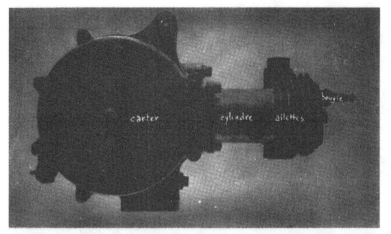

1 . moteur P.F. ancien (quatre temps,
volant dans le carter); le
carburateur a été enlevé de la
tubulure d'admission.

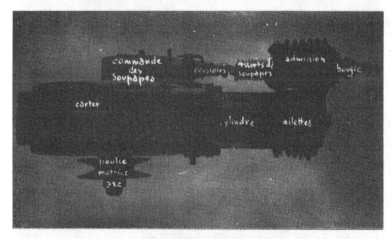

2 . moteur P.F.; noter la distance
entre la culasse et les sièges des
soupapes, de type latéral. Le point
d'allumage est loin de la culasse, ce qui
retarde l'onde explosive.

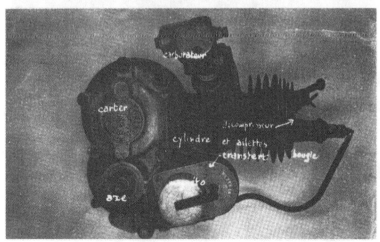

3 . moteur Zurcher (deux temps);
le carter, servant à la précompression,
est réduit. Le volant devient extérieur;
la bougie est près du point mort haut
du piston.

plate 1

cylindre Zurcher

cylindre P.F.

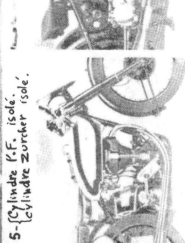

5.- {Cylindre P.F. isolé.
    {Cylindre Zürcher isolé.

8.- Moteur "Sunbeam"; le développement des ailettes à gagné le carter

7.- Moteur de motocyclette
Norton "Manx"

6.- Moteur Solex à allumage par
volant magnétique; le développement des
ailettes ne dépend pas de la puissance.

plate 2

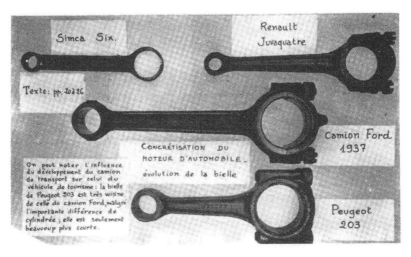

plate 3

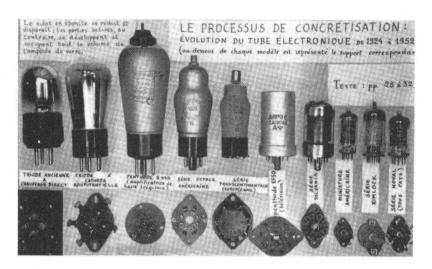

Le culot en ébonite se réduit et disparaît ; les parties actives, au contraire, se développent et occupent tout le volume de l'ampoule de verre.

LE PROCESSUS DE CONCRÉTISATION : ÉVOLUTION DU TUBE ÉLECTRONIQUE de 1924 à 1952

(au-dessous de chaque modèle est représenté le support correspondant

TEXTE : pp. 28 à 32

TRIODE ANCIENNE à CHAUFFAGE DIRECT · TRIODE à CATHODE ÉQUIPOTENTIELLE · PENTHODE E446 (amplificatrice de haute fréquence) · SÉRIE OCTALE AMÉRICAINE · SÉRIE TRANSCONTINENTALE (EUROPÉENNE) · penthode EF50 (télévision) · SÉRIE SYLVANIA · MINIATURE AMÉRICAINE · SÉRIE RIMLOCK · SÉRIE NOVAL (TOUS PAYS)

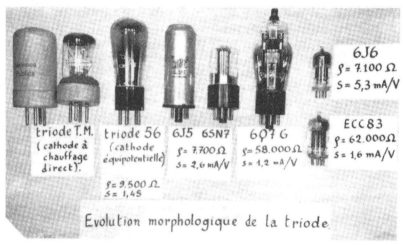

triode T.M. (cathode à chauffage direct).

triode 56 (cathode équipotentielle)
$\rho = 9.500\ \Omega$
$S = 1,45$

6J5 6SN7
$\rho = 7.700\ \Omega$
$S = 2,6\ mA/V$

6Q7 G
$\rho = 58.000\ \Omega$
$S = 1,2\ mA/V$

6J6
$\rho = 7.100\ \Omega$
$S = 5,3\ mA/V$

ECC83
$\rho = 62.000\ \Omega$
$S = 1,6\ mA/V$

Évolution morphologique de la triode.

Structures actives et structures passives

triode primitive · triode 56 · penthode E446 · duo-diode triode 6Q7G · octode EK2

plate 4

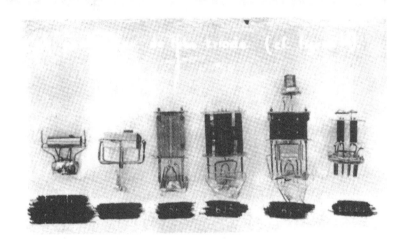

Concentration des fonctions actives sur des structures plurifonctionnelles

lampes Sylvania (1940) : blindage interne.

blindage externe

ARP3... A/3U56 A/V

EF50, penthode de télévision et de RADAR : pied en verre pressé mais blindage rapporté.

pied en verre.

14H7

partie inférieure du blindage

miniature

klimlock

noval

formes classiques actuelles.

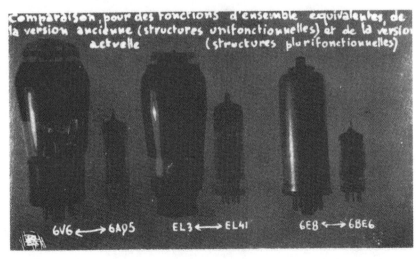

Comparaison pour des fonctions d'ensemble équivalentes de la version ancienne (structures unifonctionnelles) et de la version actuelle (structures plurifonctionnelles)

6V6 ⟶ 6AQ5   EL3 ⟵ EL41   6E8 ⟵ 6BE6

plate 5

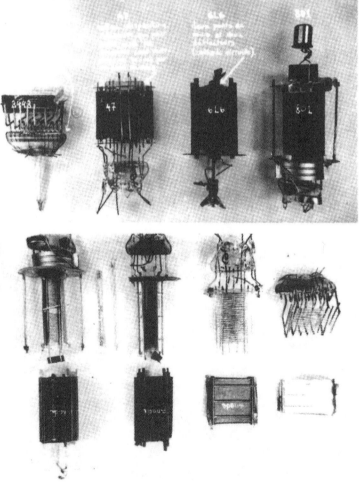

plate 6

plate 7

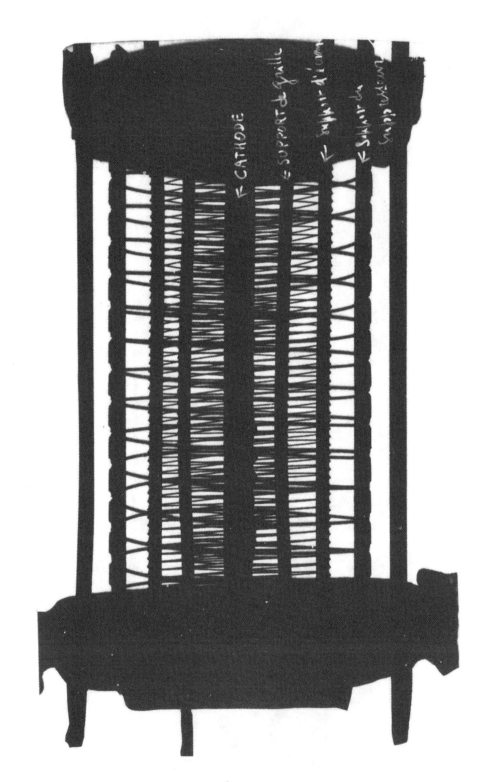

plate 8

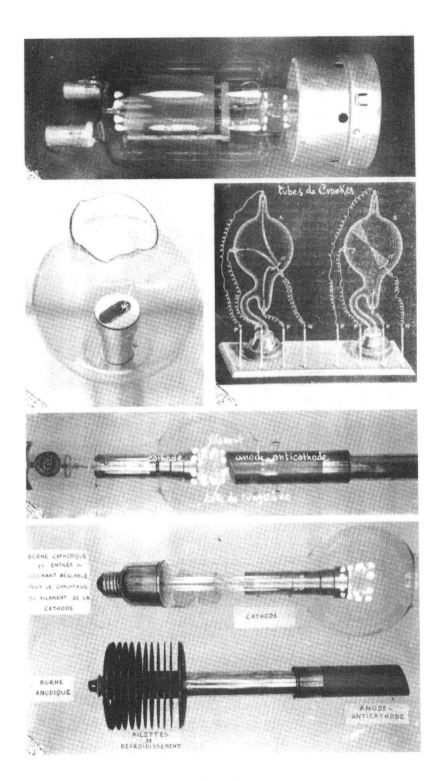

plate 9

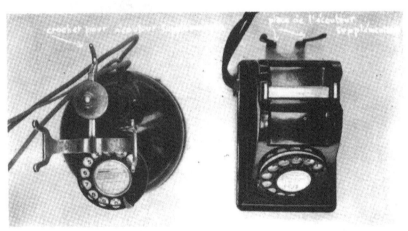

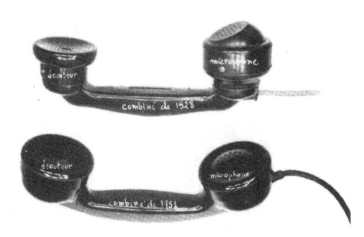

plate 10

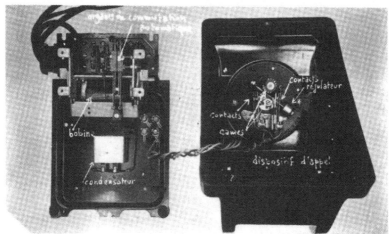

plate 11

plate 12

Agriculture, Economie Rustique

Agriculture, Economie Rustique

Agriculture, Economie Rustique

plate 13

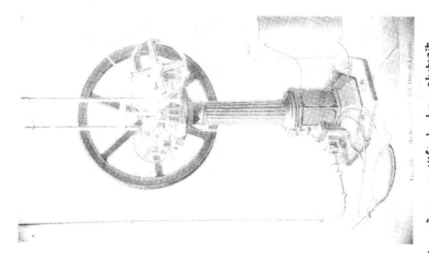

moteur à gaz primitif, de type abstrait, sans bielle ni manivelle (à crémaillère), d'après Privat-Deschanel.

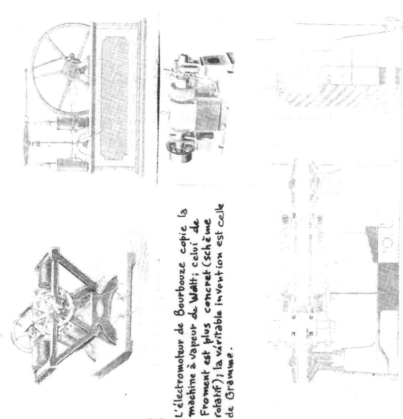

L'électromoteur de Bourbouze copie la machine à vapeur de Watt; celui de Froment est plus concret (schéme rotatif); la véritable invention est celle de Gramme.

vue perspective et coupe de la machine de Gramme, d'après Electrical Engineering

plate 14

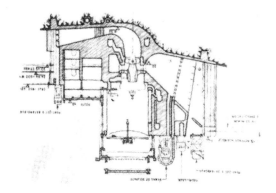

type d'usine de basse chute : coupe de l'usine de Donzère - Mondragon

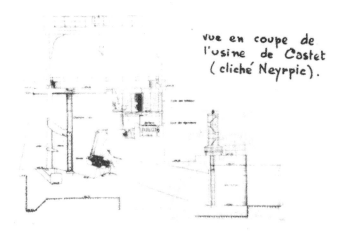

vue en coupe de l'usine de Castet (cliché Neyrpic).

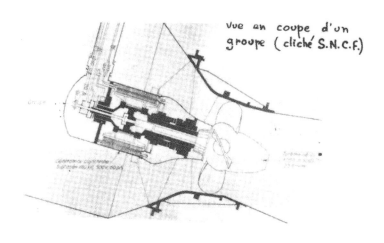

vue en coupe d'un groupe (cliché S.N.C.F.)

plate 15

PART II
Man and the Technical Object

CHAPTER ONE

THE TWO FUNDAMENTAL MODES OF RELATION BETWEEN MAN AND THE TECHNICAL GIVEN

I. – Social majority and minority of technics

We would like to show that the technical object can be connected with man in two opposing ways: according to the status of majority or of minority. The status of minority is one whereby the technical object is firstly an object of utility, necessary for everyday life, belonging to the heart of the environment where the human individual's growth and training takes place. In this case the encounter between the technical object and man occurs essentially during childhood. Technical knowledge is implicit, non-reflective, and habitual. Conversely, the status of majority corresponds to an operation of reflection and self-awareness by the free adult, who has at his disposal the means of rational knowledge, elaborated through the sciences: the knowledge of the apprentice is thus distinguished from that of the engineer. Once the apprentice has become an adult craftsman and the engineer has been integrated into a network of social relations, they retain and project a vision of the technical object that corresponds, in the first case, to the minority status and in the second case to the majority status; what we have here are two very different sources of representation and judgment related to the technical object. However, the crafts- man and the engineer don't merely live for themselves; they are both witnesses to and agents of the relation between human society as a whole and the world of technical objects, and they have an exemplary value: it is through them that the technical object is incorporated into culture. To the present day, these two modes of incorporation have not been capable of providing corroborating results, to the extent that there is something akin to two different languages and types of thought arising from technics that are incoherent with each other. This lack of coherence is partly responsible for the contradictions of contemporary culture, to the extent that it judges and represents the technical object in relation to man.

Moreover, this conflict between the majority aspect and the minority aspect is only a particular case of inadequacy that has always existed between the individual or social man and technical reality. Throughout antiquity, a large part of technical operations were excluded from the domain of thought: they were operations that corresponded to servile occupations. Just as the slave was excluded from the city, servile occupations and the technical objects corresponding to them were banned from the universe of discourse, reflexive thought, and culture. Only the Sophists, and to a certain extent Socrates, made the effort to introduce the technical operations practiced by slaves or freed men into the domain of noble thought. Majority status was granted to only a few operations, such as agriculture, hunting, war and the art of navigation. Technics that used tools were kept outside the domain of culture (Cicero takes most of his metaphors from the noble arts, in particular agriculture and navigation; the mechanical arts are rarely evoked by him).

Going further back in time one discovers that such or such a civilization also created the distinction between noble technics and non-noble technics; the history of the Jewish people reserves a privileged place for pastoral technics and considers the land as cursed. God accepts the offerings of Abel, but not those of Cain: the pastor is superior to the farmer. The Bible contains a multitude of schemas of thought and paradigms drawn from various ways for helping to foster the flock. Conversely, the Gospels introduce modes of thought taken from the experience of agriculture. Perhaps one could find a certain technological bias at the origin of mythologies and religions, consecrating one technique as noble and not granting citizenship to the others, even when they are used effectively; this initial choice between a major and a minor technique, between a valued technique and a devalued technique, grants an aspect of partiality, of non-universality to the culture that incorporates technical schemas. Our research does not propose to ascertain the reasons and modalities of this choice between fundamental techniques in each particular case, but simply to show that human thought must establish an egalitarian relation, without privilege, between technics and man. This task must still be accomplished, because these phenomena of technical dominance — because of which there is, in each epoch, a part of the technical world that is recognized by culture, while other parts of the technical world are rejected — maintain an inadequate relation between human reality and technical reality.

The disappearance of slavery in Western Europe allowed ancient servile techniques to come to light and manifest themselves in clear thought: the Renaissance recognized artisanal techniques by shedding the light of rationality onto them. Rational mechanics introduced machines into the domain of mathematical thought: Descartes calculated the transformations of movement within the simple machines used by the slaves of antiquity. This effort toward rationalization, signifying an integration into culture, continued up until the end of the eighteenth century. Yet despite this, the unity of technics did not persist; a genuine reversal took place, 126 which repressed the ancient noble techniques (those of agriculture and animal husbandry) into the domain of the irrational, the non-cultural; the relation to the natural world was lost, and the technical object became an artificial object distancing man from the world. We are only now beginning to see the possibility of an encounter between the way of thinking inspired by a technics related to living beings and an artificialist way of thinking concerned with constructing automata. Mechanical technics were only truly able to attain majority status by becoming a technics thought by the engineer, rather than remaining the technics of the craftsman; at the artisanal level, the concrete relation between the world and the technical object still exists; but the object thought by the engineer is an abstract technical object, unattached to the natural world. In order for culture to incorporate technical objects, one would have to discover an intermediary between the majority status and the minority status of technical objects. The condition of the disjunction between culture and technics resides in the disjunction that exists within the world of technics itself. In order to discover an adequate relation between man and the technical object, one would have to be able to discover a unity of the technical world, through a representation that would incorporate both that of the craftsman and that of the engineer. The representation of the craftsman is drowned in concreteness, engaged in material manipulation and sensible existence; it is dominated by its object; the representation of the engineer is one of domination; it turns the object into a bundle of measured relations, a product, a set of characteristics.

The prime condition for the incorporation of technical objects into culture would thus be for man to be neither inferior nor superior to technical objects, but rather that he would be capable of approaching and getting to know them through entertaining a relation of equality with them, that is, a reciprocity of exchanges; a social relation of sorts.   127

The compatibility or incompatibility between different technological modes is worth subjecting to a conditional analysis. Perhaps it will be possible to discover the conditions of compatibility between one technology and another, such as that of the Romans and that elaborated by the civilized societies of our day; perhaps it will be possible to discover a real, if unapparent incompatibility between the technological conditions of the nineteenth century and those of the mid-twentieth century. Certain myths born from the inappropriate encounter between two incompatible technological paradigmatisms could then be brought back to their initial conditions and analyzed.

II. – Technics learned by the child and technics thought by the adult

One cannot study the status of the technical object in a civilization without taking into account the difference between the relation of this object to the adult and the child; even if life in modern societies has accustomed us to thinking that there is continuity between the life of the child and of the adult, the history of technical education quickly shows that the distinction did exist, and that the characteristics of the acquisition of technical knowledge are not the same depending on whether this acquisition takes place in the adult or in the child; we have no intention whatsoever of laying down a normative rule, but merely of showing that the characteristics of teaching technics have varied greatly throughout time, and that they have varied not only due to the state of technics, or the structure of societies, but also because of the age of the subjects who were learning; one could discover therein a circular relation of causality between the state of technics and the age of knowledge acquisition constituting the technician's qualifications; if a very poorly rationalized technics requires an extremely precocious initiation of learning, then the subject will, even as an adult, retain a basic irrationality in his technical knowledge; he will possess it by virtue of a very profound acquisition, due to his early habitual immersion; as a result, the technician's knowledge will not consist of clearly represented schemas, but of a *manual dexterity* possessed almost by instinct, and entrusted to this second nature that is habit. His science will be at the level of sensorial and qualitative representations, very close to the concrete aspects of matter; this man will be endowed with a power of intuition and complicity with the world that will give him a very remarkable aptitude that can only manifest itself in work and not in consciousness or discourse; the craftsman will be like a magician, and his knowledge will be operational rather than intellectual; it will be an ability rather than a knowing; by its very nature, it will be a secret for others, because it will be a secret for himself, for his own consciousness.

To this day we find the existence of this technical subconsciousness, which cannot be verbalized in clear terms by reflective activity, in farmers or shepherds, capable of directly grasping the value of seeds, the exposure of a plot, the best place to plant a tree or to set up a pasture so that it is sheltered and well positioned. These men are experts in the etymological sense of the term: they take part in the living nature of the thing they know, and their knowing is a one of profound, direct participation that necessitates an original symbiosis, including a kind of fraternity with a valued and qualified aspect of the world.

Man here behaves like an animal who smells water or salt from a distance, who immediately knows where to choose the place for a nest without prior reasoning. Such participation is of an instinctive nature and it is only present when the life of successive generations has made the rhythm of life, the conditions of perception, and the essential mental structures adapt to a kind of activity aimed at a stable 129 nature. In his remarkable tale titled *The Mines of Falun*, E. T. A. Hoffmann describes a similar power of intuition in the miner; he feels danger and can find minerals in the most hidden veins; he lives in a kind of co-natural relation with underground nature, and this co-naturalness is so profound that it excludes all other feeling or attachment; the true miner is a subterranean man; the one who descends into the mine without loving it won't discover this essential co-naturalness, as is the case with the errant sailor who bravely signs up to work in the mine because he loves a young girl; he becomes the mine's victim, on the very morning of his wedding. There is no morality here; the young sailor is full of merit and virtue. But he is a sailor, and not a miner; he doesn't have the intuition of the mine. The phantom of the old miner warns him of the danger he is in, because the mine rejects the intruder, as one who comes from outside, from another trade, from another life, and who is not endowed with the power of participation. In the farmer, the shep-herd, the miner, and the sailor, human nature doubles as a second nature which is like an ancestral pact with an element or a region. It is difficult to say whether this sense of participation is acquired in the first years of life or whether it is involved in a hereditary patrimony; what remains certain is that this kind of technical training [*formation*], consisting of intuitions and purely concrete operative schemas that are very difficult to formulate or transmit through any kind of symbolism, be it oral or figurative, belongs to childhood. For this same reason it is unlikely to evolve, and can hardly be re-formed in adulthood: it is not, in fact, of a conceptual or scientific nature, and cannot be modified through oral or written symbolism.

130   This technical training is rigid. It would be entirely wrong to consider this technical training as necessarily inferior to a training using intellectual symbols; the quantity of information of this type of instinctive training can be as great as that contained within knowledge clearly explained through symbols, graphs, schemas, or formulas; it is too easy to oppose routine to science, which would by the same token be progress; primitiveness cannot be confused with stupidity, any more than conceptualization with science. But it is worth noting that this technical knowledge is effectively rigid, since man cannot become a child again in order to acquire new basic intuitions. Furthermore, this form of technics has a second characteristic: this characteristic is one of initiation and it is exclusive; it is indeed by growing up in a community that is already fully saturated with the schemas of a particular work that the child acquires base intuitions; the one who comes from the outside will more than likely be deprived of this initial participation that requires the existence of vital conditions, because the vital conditions are educational in this primary sense. It would, without a doubt, be wrong to attribute the closure of ancient technics to the closure of the communitarian life of societies: such societies in fact knew how to open themselves up, as the temporary or seasonal migration of farmers from Auvergne to Paris toward the end of the nineteenth century shows; it is technics itself that corresponds in this case to a closed regime of life, because technical training is valid only for the society that formed it, and is therefore the only one valid for that society. It seems that historians are inclined to consider the rites of initiation of ancient trades in a rather abstract manner, by looking at them from a purely sociological point of view; it is worth noting that these are tests that correspond to a regime of acquisition of technical knowledge by the child; the test is thus not only a social rite, but also an act through which the young subject

131 becomes an adult by taming the world, by measuring himself with it in a critical circumstance and by triumphing. There is a certain magic in this test, which is an act through which the child becomes a man, by using all his strength pushed to its extreme limit for the first time. If he weakens or shows himself to be inferior in this dangerous confrontation with the world and with matter, he risks foregoing the effectiveness of his manly action. If a hostile nature cannot be defeated, the man cannot become a complete adult, because a gap has arisen between nature and himself; the test is a life-long enchantment of the technical being; it is an operation that creates the obedience of matter to man, who has become its master because he succeeded in taming it, like an animal that becomes docile from the day it first allowed itself to be led. If the first gesture is a mishap, the animal rebels and remains wild; it will never again accept this master, who will, himself, forever be lacking

in self-assurance, because the immediate contact has been broken. In this test, a law of all or nothing manifests itself; man and the world are therein transformed; an asymmetrical union is constituted; one must not say that the test is a demonstration of courage or ability as if it were simply an exam; it creates these qualities, because courage is made by way of an immediate and certain connection with the world, which wards off all uncertainty and all hesitation; courage is not a defeated fear, but a fear always warded off by the presence of an intuition because of which the world is with the one who acts; the skillful man is the one whom the world accepts, whom matter loves and obeys with the faithful docility of an animal who has recognized his master. Skill is one of the forms of power, and power supposes an enchantment making an exchange of forces possible, or rather a more primitive and more natural mode of participation than that of enchantment, which is already very elaborate and partially abstract. In this sense, skill is not the exercise of a violent despotism, but of a force conforming to the being who conducts it. There is, in the true power of the skillful man, a relation of recurrent causality. The   132 true technician loves the matter upon which he acts; he is on its side, he is initiated but respects that to which he is initiated; he forms a couple with this matter, after having tamed it, and only delivers it with caution to the profane, because he has a sense of the sacred. The craftsman and the farmer to this day still experience an aversion to consigning some works or products to commerce that express their most refined and most perfect technical activity: this prohibition of commerciality, of divulgation, can be seen in the not-for-sale prints of a book that a printer, editor, or author can distribute. It can also be seen in the Pyrenean farmer who offers his visitor a certain food in his home, which can neither be bought nor taken away.

The secret and stable aspect of such a technics is thus not only a product of social conditions; it produces the structure of groups as much as this structure of the group conditions it. It is possible that every technics must to a certain extent contain a coefficient of intuition and instinct necessary for establishing the appropriate communication between man and the technical being. But besides this first aspect of technical formation, there exists a second aspect that stands in an inverse relation to the first, and which is essentially aimed at the adult. As in the preceding case, there is a dynamic action exerted on the individual man and on the group, leading him to possess an adult mentality.

This second type of technical knowledge is a rational, theoretical, scientific, and universal knowledge. The best example is Diderot and d'Alembert's *Encyclopedia*. If the *Encyclopedia* appeared as a powerful and dangerous work, then it was not because of its veiled or direct attacks on a certain abuse of privilege, nor because

of the "philosophical" aspect of certain articles; there were more violent libels and
pamphlets than the *Encyclopedia*. The *Encyclopedia* was respectfully feared because
it was moved by an enormous force, that of a technical encyclopedism, a force
that had brought together powerful and enlightened protectors; this force existed
by itself, because it responded to the needs of its time, more than political or
financial reforms did; it was this force that was positive and creative, and which
realized an equally remarkable assembly of researchers, editors, and correspondents
by granting a faith to this team composed of men who collaborated without being
connected through social or religious communities; a great work had to be carried
out. The greatness of the *Encyclopedia*, its novelty, resides in the fact that its prints
of schemas and models of machines, which are an homage to the trades and to the
rational knowledge of technical operations, are fundamentally major. But these
prints do not have the role of pure, disinterested documentation for a public eager
to satisfy its curiosity; the information in them is complete enough to constitute a
useable practical documentation, such that anyone who owns the book would be
capable of building the described machine or of further advancing the state reached
by technics in that domain through an invention, and to begin his research where
that of others who preceded him leaves off.

The method and structure of this new way of teaching stands in an inverse relation to the preceding one: it is rational and doubly universal; this is why it is adult.
It is rational because it employs measurement, calculation, procedures of geometrical figuration and descriptive analysis; it is also rational because it calls upon
objective explanations and invokes experimental results, with the aim of a precise
presentation of conditions, treating as a hypothesis whatever is conjectural and as
established fact whatever one must consider as such; not only is scientific explana-
tion required, but it is required with a clear taste for the scientific spirit. Moreover,
this way of teaching is doubly universal, both through the public it addresses and
through the information it provides. Of course, what is taught here is high-level
knowledge, but despite this, it is meant for all; the cost of the book alone limits
the possible purchases. This knowledge is given in the spirit of the highest possible
universality, according to a circular schema that never supposes that a technical
operation is closed in on itself through the secret of its specialty, but rather that
it is related to other operations, using analogous types of apparatuses, and based
upon a small number of principles. For the first time, one sees a technical universe
constituting itself, a cosmos wherein everything is related to everything else rather
than being jealously guarded by a guild. This consistent and objective universality,
which supposes an internal resonance of this technical world, requires that the

book be open to all and that it constitute a material and intellectual universality, a block of available and open technical knowledge. This teaching supposes an adult subject, capable of directing himself and of discovering by himself his own normativity without being directed by another being: the autodidact is necessarily an adult. A society of autodidacts cannot accept tutelage and the status of spiritual minority. It aspires to govern itself on its own, and to manage itself. It is principally in this sense and through its technological power that the *Encyclopedia* brought about a new force and a new social dynamic. The causal circularity of encyclopedic knowledge excludes the moral and political heteronomy of the monarchy [*l'Ancien Régime*]. The technical world discovers its independence when it realizes its unity; the *Encyclopedia* is a kind of Fête de la Fédération[1] of technics discovering their solidarity for the first time.

III. – The common nature of minor technics and major technics.
The signification of encyclopedism   135

We shall attempt to analyze the relation between the encyclopedic spirit and the technical object, because it appears indeed to be one of the poles of all technical awareness, and thus possesses, in addition to its historical signification, a sense of the knowledge of technicity that is still valid. We have opposed the implicit, instinctive, and magical aspect of technical education aimed at the child to the inverse aspects of the latter, which one discovers in the *Encyclopedia*; but this opposition runs the risk of masking a deep analogy of the dynamisms inherent in these structurations of technical knowledge; encyclopedism manifests and propagates a certain inversion of technics' fundamental dynamisms; this inversion is nevertheless possible only because operations are not annihilated, but displaced and in a way turned around. The *Encyclopedia* also manipulates and transfers forces and powers; it too performs an enchantment and draws a circle like the magic circle; except that it does not enchant by the same means as those inherent in the testing of instinctive knowledge, and it is not the same reality that it places within the circle of knowledge. It is human society with its forces and obscure powers that is placed within the circle, having become immense and capable of comprising everything. This circle is represented and constituted by the objective reality of

---

1.   The *Fête de la Fédération* was the massive national celebration, held on July 14th 1790, to mark the first anniversary of the storming of the Bastille and during which every citizen, including the king, was to take an oath of fidelity to the nation and to the new constitutional order, that was held in order to honor what was thought at the time to be the successful fulfillment of the Revolution and to celebrate the peaceful and triumphant unity of the different elements of the nation on the basis of new, rational principles. [TN]

the book. Everything that is represented in the encyclopedic book is at the service of the individual who has in his possession a figural symbol of all human activities in their most secret details. The *Encyclopedia* makes initiation universal, and thereby produces a sort of rupture in the very sense of initiation; the secret of the objectified universal maintains a positive sense of the notion of the secret (the perfection of knowledge, a familiarity with the sacred), but annihilates the negative

136   aspect (obscurity, a means of exclusion through mystery, knowledge reserved to a small number of men). Technics becomes an exoteric mystery.[2] The *Encyclopedia* is a magic cipher [*voult*][3] and is all the more efficient as it has been built with a more precise, more exact and more objective representation of its model; all the active resources, all the living forces of human operations are assembled in this object-symbol. Each individual capable of reading and of understanding possesses the *voult* of the world and of society. Magically, everyone is master of everything, because he possesses the *voult* of the whole. The cosmos, once enveloping and superior to the individual, and the social circle constraining and always eccentric with respect to the power of the individual, are now in the hands of the individual, like the globe representing the world which the emperors carry as a sign of sovereignty. The power, the confidence of the reader of the encyclopedia is the same as that of a man who first attacked the effigy of an animal before confronting it in nature, the same once again as that of the primitive farmer who entrusted the seeds to the soil after having performed propitiatory rites, or of the voyager who ventured to new places only after having rendered them in some way favorable through an act of establishing communion and pre-possession of which the Odyssey preserves a memory.[4] The gesture of initiation is a union with a reality that remains hostile so long as it hasn't been tamed and possessed. It is for this reason that all initiation leads to virility and adulthood. Thus from a psycho-sociological point of view, every manifestation of the encyclopedic spirit can appear, within a society, as a fundamental movement [*mouvement de fond*] expressing the need for attaining a state

137   of freedom and adulthood, since the current regime or customs of thought retain individuals within a state of tutelage and artificial minority; this will to move from minority to majority by way of enlarging the circle of knowledge and liberating the power inherent in knowing, is what we encounter on three occasions in the

---

2.   Part of this feeling of the efficacy of primitive magic has turned into the unconditional belief in progress. The object that is modern or has a modern aura [*allure*] is endowed with an almost supernatural [*surnaturel*] efficacy. The feeling of modernity comprises something of the belief in an unlimited and polyvalent power of a privileged object.

3.   *Voult:* in English, 1. Poppet, wax or clay image or doll (poppet) of a person used in witchcraft or voodoo to affect him magically; 2. Old word for face, for instance work representing the face of Christ. Generally, in Simondon's usage, a symbol or *analogon* of a certain reality, in the form of an object, an image, or a piece of an image in which the part stands for the whole, and by means of which the reality that is symbolized comes into the power of the one who possesses it, as when a spell is cast [*envoûtement*]. [TN]

4.   The rite of possession of the earth accomplished by Ulysses approaching the island of the *Phaeacians*.

history of thought since the Middle Ages. The first manifestation of encyclopedic spirit is what constitutes the Renaissance and is contemporary with the ethical and religious revolution of the Reformation. To want to move from the Vulgate to the veritable text of the Bible, looking for the Greek texts rather than contenting oneself with poor Latin translations, rediscovering Plato beyond a scholastic tradition crystallized according to a fixed dogma, is to refuse the arbitrary limitation of thought and of knowledge. Erudition represents not a return to the past as past, but the will to enlarge the circle of knowledge, to rediscover all of human thought in order to be freed from a limitation of knowledge.

The humanism of the Renaissance is not at all a will to redeploy a fixed image of man in order to restrain and normalize knowledge, as the decay into which the study of classical antiquity has fallen would appear to suggest today. Humanism first of all responds to an encyclopedic movement [*élan*]. But this movement is always oriented toward already formalized knowledge because the level of technical development was not sufficiently advanced so as to allow a rapid formalization of this domain to intervene; The sciences in particular were too underdeveloped; the intellectual means of universalizing technics were not ready; it is the seventeenth century that brought about the means for universalizing technics which the Encyclopedia then put to work; nevertheless, we should note the very positive attitude toward technics from the Renaissance onward; already with the Renaissance, technics are valued as a paradigm and means of expression,[5] or for their human value which opens new paths. Rabelais's magnificent praise of the Pantagruelion plant[6] summarizes all the hopes of the men of the Renaissance, all their beliefs in the "virtue" of technics, thanks to which humanity would perhaps one day be   138
capable of traveling all the way "up to the celestial signs," in the same way that humanity had travelled from the Old World to the New World.

The second encyclopedic stage is that of the Enlightenment; scientific thought was freed, but technical thought did not find liberation; it is scientific thought that freed technical thought. Since technics touch upon commerce, agriculture, and industry, and these are aspects of society, this technological encyclopedism could not help but be correlated with social and administrative reforms. Institutions

---

5.   In *Defense and Illustration of the French Language* (Joachim du Bellay). Rabelais and Montaigne also employ many terms drawn from crafts and trades.

6   Pantagruelion is an herb with numerous marvelous properties, similar to hemp, discovered by the character Pantagruel, which Rabelais describes extensively in the final chapters of *The Third Book of Pantagruel*; see *The Complete Works of François Rabelais*, trans. Donald Frame (Berkeley, CA: University of California Press, 1991). [TN]

such as the elite schools [*Grandes Écoles*] emerge from this encyclopedism; encyclopedism is by definition polytechnic in its industrial form, just as it is physiocratic in its agricultural form. The industrial aspect becomes more developed than the physiocratic aspect, because encyclopedic rationalization facilitated more pronounced transformations in the industrial domain, benefiting from the recent scientific discoveries of the late eighteenth century. This asymmetrical development, however, should not make us forget that one of the most important components of the technical encyclopedic spirit is the direct link between the individual with the vegetal and animal world, with biological nature; rather than leaving it to the descendants of former serfs, the art of plowing is valued even by the most distinguished figures. This is the "pastoral" era, and a time where a mind as solid as that of Daubenton sees no shame in writing a treatise for shepherds which is the prototype of generous, high level popularization, unifying the old tradition of didactic works and giving it a new lease on life through the use of a clear graphic symbolism that is almost comprehensible enough for the illiterate; the substance of this beautiful book resides in its etchings, as neat and expressive as those of the *Encyclopedia*.

139 It must indeed be noted that technology calls for a means of expression other than oral expression, which uses already known concepts, and which can transmit emotions, but struggles to express schemas of movement or precise material structures; a symbolism adequate for technical operation is visual symbolism, with its rich play of forms and proportions. The civilization of the word gives way to that of the image. The civilization of the word, in turn, is by its very nature even more exclusive than that of the image, because the image is by nature universal, requiring no prior code of significations. All verbal expression tends to become initiatory; it becomes specialized by becoming a kind of ciphered language, of which the old jargon of the guilds is a clear example. One has to belong to a closed group in order to understand the oral or written language; to understand schematic expression, it is enough to be able to perceive. It is with the schema that technical encyclopedism takes on all its meaning [*sens*][7] and its power of diffusion, by becoming truly universal. The printing press had given rise to a first phase of encyclopedism by distributing texts; but this encyclopedism could only reach reflexive or emotive significations, already sanctioned by the constituted culture; with the printed word as intermediary, the information going from individual to individual makes a detour through the social institution that is language. Through the mediation of the visual sign, printed writing first transmits an oral message, with all the limitations inherent in this mode of expression; to be in possession of all the modern and all the

---

7. "*Sens*" in French signifies both "meaning" and "direction toward which," as well as "sense." [TN]

dead languages is necessary for the understanding of an encyclopedism of verbal significations; this possession, or at least the effort going toward this possession, is in part the meaning [*sens*] of the Renaissance, but in fact it remains the privilege of humanists and scholars; culture does not have any direct universality through oral or written language. It is perhaps for this reason that the Renaissance was unable to establish a technological universality, even though it had a tendency to prefer sculptural and graphic expression to all other symbolism, especially in the   140 arts. Printing, which is a faculty for the diffusion of a spatial schema, acquires its full sense in etchings. Yet, symbolic etching, used as a means for clearly translating a thought into the terms of structures and operations, freed of any intention in the direction of allegorical expression returning to oral expression (as on a coat of arms), appears with its complete development in the seventeenth century, as for instance in Descartes's treatises. Having borrowed its expressive force and its power of precision from the common use of geometry, it was now ready to constitute the adequate symbolism of a universal technology.

A third stage of encyclopedic thinking finally appears to announce itself in our own era, but hasn't yet succeeded in constituting its modes of universal expression. The civilization of oral symbolism has once more overcome that of spatial, visual symbolism, because the new means of diffusion of information have given primacy to oral expression. When information must be converted into a printed object and transported, the delay separating the discovered thought from the expressed thought is the same for written information as it is for figural information. Printing tends to privilege figural information, because it necessarily uses the spatial form; the schema is that which does not require translation into a form other than its original form, whereas writing is the translation of a series that is temporal in its origin into a spatial series, which will have to be converted back into a temporal series upon reading it. On the contrary, in the case of information transmitted by telephone, telegraph or radio waves, the means of transmission requires the translation of a spatial schema into a temporal series, and subsequently its conversion back into a spatial schema; radio diffusion in particular is directly suited to oral expression, and may be adapted to the transmission of a spatial schema only with great difficulty; it consecrates the primacy of sound. Spatial information is thus relegated to the domain of expensive or rare things, always late with respect to oral   141 information, which is valued because it follows every step of vital[8] coming-into-being. A civilization, however, is guided by a latent paradigmatism at the level of the information it values; this paradigmatism has once again become oral; thinking

---

8.   Or social.

once again takes place according to verbal semantemes, of the order of the slogan. The acting presence of inter-human relations is of the order of the verb. There is of course cinema and also television. But we must note that, due to the very dynamism of its images, cinematography is a cinematic, dramatic action, rather than a graphics of simultaneity, and not a direct expression of the intelligible and stable form; subsequent in its discovery to the first attempts at the transmission of images by television, cinematography has completely supplanted the latter and has imposed on it the dynamism of images, which still burdens television with an enormous task, and turns it into a competitor or imitator of cinematography, incapable of discovering its own modes of expression, enslaved to the public as a means of pleasure. Cinematographic movement is rich in hypnosis and rhythm that dulls the reflexive faculties of the individual in order to induce a state of aesthetic participation. Organized according to a temporal series that employs visual terms, cinema is an art and a means of expressing emotions; the image here is a word or a phrase, it is not an *object* comprising a structure to be analyzed by the activity of the individual being; it rarely becomes an immobile and radiating symbol. Furthermore, television could become a means of information contemporary to human activity, which cinema cannot be, since, being a fixed and recorded thing, it puts everything it incorporates into the past. But since television wants to be dynamic, it is obliged to transform every point of each image into a temporal series, in as short a time as the projection of each static image in the cinema. Therefore, it must first of all transform the dynamic into the static, through a first cut into images. Then, during the transmission of each fixed image, it transforms the simultaneous points of this fixed image into a temporal series; upon its arrival, each temporal series transforms itself into a spatial and immobile picture, and the rapid succession of these fixed images, as in cinematography, recreates the analyzed movement, according to the characteristics of the perception of movement. This double transformation consequently results in the necessity of transmitting an enormous quantity of information, even for an image that is extremely simple in its intelligible structure. There is no common measure between the quantity of information that is effectively interesting and significant for the subject, and the quantity of information that is technically employed, corresponding to several million signals per second. This waste of information hinders television from being a subtle and faithful means of expression for the individual, and prevents a veritable visual symbolism from constituting itself universally; radio broadcasts cross boundaries whereas visual information often remains tied to the communitarian life of groups; it cannot be valued in these conditions. But research into coding systems, useful for inscribing the results of calculating machines on a cathode ray oscilloscope screen, or for displaying signals of electromagnetic[9] detection on the same type of screen, appear capable of conveying a very great simplification of the Hertzian transmission of schematic images; visual information would thus regain the place that it has lost because of radio broadcasts with respect to spoken language and would be capable of giving rise to a new universal symbolism.

The encyclopedic intention, in turn, begins to show itself within the sciences and technics, through the tendency toward rationalization of the machine and through the establishment of a symbolism common to the machine and to man; it is because of this symbolism that a synergy between man and machine is possible; for common action requires a means of communication. And since man cannot have several types of thought (every translation corresponds to a loss of information), it is this mix of the relation between man and machine that a new universal symbolism must emulate in order to be homogeneous with a universal encyclopedism.   143

Cybernetic thinking has already led in information theory to research such as that into *human engineering*,[10] which specifically studies the relation between man and machine; one can henceforth conceive of an encyclopedism on a technological basis.

This new encyclopedism must, like the two preceding ones, bring about a liberation, but in a different sense; it cannot be a repetition of the Enlightenment. In the sixteenth century man was enslaved to intellectual stereotypes; in the eighteenth century, he was limited by the hierarchical aspects of social rigidity; in the twentieth century, he is enslaved to his dependence on unknown and distant powers that direct him while he can neither know nor react against them; it is isolation that enslaves him, and the lack of homogeneity of information that alienates him. Having become a machine in a mechanized world,[11] he can regain his freedom only by taking on this role and by superseding it through an understanding of technical functions thought from the point of view of their universality. Every encyclopedism is a humanism, if by humanism one means the will to return the status of freedom to what has been alienated in man, so that nothing human should be foreign to man; however, this rediscovery can take place in different ways, and   144

---

9.   In particular in R.A.D.A.R., Radio Detection and Ranging (detection and measure of distance through radio waves).

10.   English in original. [TN]

11.   Today man follows a strong tendency that drives him to behave like a tool-bearing machine, because for many centuries prior to the creation of machines he fulfilled this function, at a time when technical elements existed, in the form of tools, and technical ensembles existed in the form of workshops or building sites, but not as technical individuals in the form of machines.

each age recreates a humanism that is to a certain extent always appropriate to its circumstances, because it takes aim at the most severe aspect of alienation that a civilization contains or produces.

The Renaissance defined a humanism capable of compensating for the alienation resulting from ethical and intellectual dogmatism; it aimed at regaining the freedom of theoretical intellectual thought; the eighteenth century wanted to rediscover the signification of the effort of human thought applied to technics, and with the idea of progress, rediscovered the nobility of creative continuity that can be found in inventions; it has defined the right of a technical initiative to exist despite the inhibiting forces of societies. The twentieth century seeks a humanism capable of compensating for the form of alienation that intervenes within the very development of technics, through a series of specializations that society demands and produces. There appears to be a singular law of the transformation [*devenir*] of human thought, according to which any ethical, technical, and scientific invention, which sets out as a means of liberation and rediscovery of man, becomes through its historical evolution an instrument that turns against its liberation and enslaves man by limiting him: at its origin Christianity was a liberating force, calling on man to go beyond the formalism of customs and the institutional prestige of ancient society.

It was the thought according to which the Sabbath is made for man, and not man for the Sabbath; it is this same Christianity, however, which the reformers of the Renaissance accused of being a force of rigidity, tied to a constraining formalism and dogmatism, contrary to the real and profound sense of human life. The Renaissance opposed Physis to Antiphysis. Similarly, technics, invoked as a liberation through progress during the Enlightenment, are today accused of oppressing man and of reducing him to slavery by denaturing him, of estranging him from himself through a specialization that is a barrier and source of incomprehension. The center of convergence has become a principle of partitioning. This is why humanism can never be a doctrine or even an attitude capable of being defined once and for all; each epoch must discover its humanism, by orienting itself toward the main danger of alienation. During the Renaissance, the rigidity of dogma led to the emergence of a new fervor and a new movement [*élan*].

In the eighteenth century, the infinite fragmentation of social hierarchy and closed communities pushed toward the discovery of a means of universal and unmediated efficacy, overcoming, by way of the rationalization and universalization of the technical gesture, all barriers and prohibitions that custom had established. In the twentieth century, it is no longer the hierarchical or local fragmentation of

society that creates the alienation of human society with respect to man, but rather its vertiginous, unlimited and moving immensity; the human world of technical action has once again become a stranger to the individual through its development and formalization, hardening itself into a form of machinism that has now become a new attachment of the individual to an industrial world that exceeds the dimension and possibility of thinking the individual. The liberating technics of the eighteenth century is at a human scale because it is of the artisanal type. The technics of the twentieth century is beyond the forces of the individual, and constitutes a compact and resistant, but alienated human reality within the industrial world, completely beyond the grasp of the individual just as it was for the previously   146
hierarchized society.

Man no longer needs a universalizing liberation, but a mediation. The new magic will not be found in a direct expression of the individual power to act, assured by the knowledge that gives each gesture effective certainty, but in the rationalization of forces that situate man by giving him meaning within a human and natural ensemble. The very fact that teleology is treated as a knowable mechanism that is not definitively mysterious is indicative of the attempt not to accept a situation as one simply lives it and is subjected to it. Rather than seeking the procedure for the fabrication of objects without making a pact with matter, man frees himself from his situation of being enslaved by the finality of the whole, by learning how to create finality, by learning how to organize a finalized whole that he judges and appreciates, so as not to have to be passively subjected to a *de facto* integration. Cybernetics, being a theory of information and as a consequence also a theory of finalized structures and dynamisms, frees man from the constraining closure of organization by enabling him to judge this organization, rather than being subjected to it while venerating and respecting it because he is not capable of thinking or constituting it.[12] Man overcomes enslavement by consciously orga-   147
nizing finality, just as he dominated the unfortunate necessity of work during the eighteenth century by rationalizing it so as to render work efficient rather than suffering through it in resignation. Human society, in knowing its own teleological mechanisms, is the result of conscious human thought, and consequently incorporates those who create it; it is the product of a human effort of organization, and

---

12. In past centuries, an important cause of alienation lay in the fact that the human being lent his biological individuality to technical organization: he was bearer of tools; technical ensembles could establish themselves only by incorporating man as tool bearer. The deforming aspect of the profession was at once psychic and somatic. The tool bearer was deformed by the usage of tools. Professional somatic deformations have become rare in the present day. In the repugnance felt by the gentleman [*l'honnête homme*] toward men of the trade, there is perhaps a part of the unpleasant feeling that one has when seeing a monstrosity. Today's occupational *hazards* are minimal with respect to professional deformations of the past. For Plato the βάναυσος [*banausus*] is a bald dwarf. In the sung legend, the shoemaker is a disinherited creature.

creates an adequation between the fact of being situated and the fact of situating oneself. The place man has in a society thereby becomes a relation between an element of activity and an element of passivity, as a mixed status always liable to being taken up again and improved, because it is something human that is interrupted but not alienated. Consciousness is at once a demiurgic activity and the result of a preceding organization; social reality is contemporary with human effort and homogeneous with it. Only a schema of simultaneity, a constellation of forces represented in their relational power, can be adequate for this type of reality. Its development is what such a dynamic representation of man in society postulates; cybernetic schemas can only acquire a universal sense in a society that is already constituted in a manner that conforms to this thought; the most difficult reactivity to establish is that of a society in relation to cybernetic thought itself; it can constitute itself only progressively and via the mediation of already established channels of information, such as the exchanges, for example, between technics working synergistically on a given point; it is this type of grouping that Norbert Wiener cites as the source of this new technology, which is a technics of technics, at the beginning of his *Cybernetics,* published in 1948 and which is a new *Discourse on Method,* written by a mathematician teaching at an institute of technology. Cybernetics grants man a new type of majority, one that penetrates the relations of authority by distributing itself across the social body, and discovers the maturity of reflection beyond the maturity of reason, thereby giving man, in addition to the freedom to act, the power to create organization by establishing teleology. Consequently both finality and organization, which can now be rationally thought and created since they become a matter of technics, are no longer ulterior, superior reasons, capable of justifying everything: if finality becomes an object of technics, then there is something beyond finality in ethics; Cybernetics, in this sense, frees man from the unconditional prestige of the idea of finality. Man freed himself, through technics, from social constraint; through the technology of information, he becomes creator of this organization of solidarity that hitherto imprisoned him; the stage of *technical encyclopedism* can only be provisional; it calls for a stage of a *technological encyclopedism* that completes it by giving the individual a possibility of returning to the social, which now changes status, and becomes the object of an organizational construction, rather than being the acceptance of a valorized given or one that is fought, but which subsists with its primitive characteristics, external to the activity of man. Individual nature is thus no longer external to the human domain. After acceding to freedom comes access to authority, in the full sense of the term, i.e., that of creative force.

Such are the three stages of the encyclopedic spirit, which was first ethical, then technical, and which can then become technological by going beyond the idea of finality taken as ultimate justification.

However, it is wrong to say that the technics of finalized organization are useful only because of their practical results; they are useful in the sense that they bring finality from the magical level to the technical level. Whereas the evocation of a superior end, and of the order that realizes this end, is considered to be the final term in the search for its justification (because life is conflated with finality, in an age when technical schemas are mere schemas of causality), the introduction of technological schemas of finality in thought plays a cathartic role. That of which   149 there is a technics cannot act as an ultimate justification. Both individual life and social life contain many aspects of finalized processes, but perhaps finality is not the most profound aspect of individual or social life, any more so than the different modalities of finalized actions, such as adaptation to a milieu.

One could undoubtedly say that it is not a veritable finality that animates the processes of recurrent causality with negative reaction; at the very least this technical production of teleological mechanisms enables the most inferior, most primitive aspect of finality to leave the magical domain behind: the subordination of a means to an end, hence the superiority of an end with respect to its means. By becoming a technical matter, such organization is henceforth only one of the aspects of social or individual life, and its prestige can no longer mask the possibilities for the development, advent, and emergence of new forms, which cannot be justified by finality, since they produce their own end as the last term of evolution; evolution *maladapts* as much as it adapts. The realization of adaptations is but one of life's aspects; homeostases are partial functions; technology, in incorporating them and allowing them not only to be thought, but to be brought into existence rationally, leaves the open processes of social and individual life fully exposed. In this sense, technology reduces alienation.

IV. – Necessity of a synthesis between the major and minor modes of access to technics in the domain of education

The separation between the education of the adult and the education of the child in the area of technology corresponds to a difference in structure between the   150 two normative systems, and, to a certain extent, a difference between the results.

As a consequence a gap remains between pedagogical technology and encyclopedic technology that has yet to be bridged.

Encyclopedic technological education aims at giving the adult the feeling that he is a fulfilled, entirely realized being, in full possession of his means and his forces, an image of the individual man in his state of real maturity; the necessary condition of this feeling is the universality of knowledge in theory and in practice; and yet within encyclopedic learning there remains something that is abstract, and an irreducible lack of universality: the material combination of all technical devices within a technological volume that assembles and coordinates them according to an order of simultaneity or reason neglects the temporal, successive, quantic aspect of the discoveries that have led to the current state; one grasps all at once and in actuality what is progressively constructed, what is slowly and successively elaborated; the idea of progress, or rather what is mythical about it, comes from this illusion of simultaneity, which presents as a fixed state what is merely a stage; by excluding historicity, encyclopedism introduces man to the possession of a false entelechy, because this stage is still rich in virtuality; there is no determinism that presides over invention, and where progress is thought as being continuous, it masks the very reality of invention. The autodidact is tempted to bring everything back to the present: the past, insofar as he assembles it within his present knowledge, and the future, to the extent that he considers it as necessarily flowing in a continuous manner from the present through the intermediary of progress. What the autodidact lacks is having been raised in a progressive way, which is to say, having become an adult through a temporal series of developments that are structured by crises determining and enabling the passage to another phase; the historicity of technical coming-into-being must be grasped through the historicity of the subject's coming-into-being so as to add the order of succession to the order of simultaneity, according to a form that is time. A genuine encyclopedism, which demands a temporal universality at the same time as a universality of simultaneity, must integrate the education of the child; it can only become truly universal if it makes an adult by means of the child, by following the temporal universality so as to obtain a universality of simultaneity; what must be discovered is the continuity between the two forms of universality.

Inversely, non-technological education lacks the universality of simultaneity, this is what is meant when one says that it sets its sights on culture rather than knowledge; but any enterprise that aims to obtain culture by ridding itself of knowledge would be illusory, because the encyclopedic order of knowledge is a part of culture; and yet, the encyclopedic order of knowledge can only be seen in an abstract and consequently non-cultural way, if it is seen from outside of knowledge itself.

A representation of knowledge without knowledge itself can only occur if it is grasped through an external symbol, as is the case for instance in the mythical and socialized representation of men who "embody" knowledge: knowledge itself is then replaced by the figure of the scholar, which is to say by a cataloged element of social or characterological typology, which is totally inadequate to knowledge itself, and introduces a mystification into culture making it inauthentic. At best, knowledge can be replaced by an opinion, a biography, a character trait or a description of the personality of a scholar; but these are once again totally inadequate elements, because they do not introduce knowledge but rather an idolatry of the human basis of knowledge, which is not of the order of knowledge itself. There is more authentic culture in the gesture of a child who reinvents a technical device, than in a text where Chateaubriand describes the "terrifying genius" of Blaise Pascal. We are closer to invention when we seek to understand the cog-wheeled adding device used in Pascal's calculating machine (arithmetic machine) than when we   152 read the most oratorical passages relating to Pascal's genius. To understand Pascal is to reconstruct a machine identical to his with one's own hands without copying it, even transposing it where possible to an electronic adding device, so as to have to reinvent it by way of actualizing it, rather than reproducing Pascal's intellectual and operational schemas. To cultivate oneself is to actualize real human schemas analogically, paying scant attention to the stir that this or that invention or publication caused among its contemporaries, which is inessential, or at the very least cannot be grasped other than with reference to an original thought, to invention itself.

It is regrettable that a cultured pupil in the last year of secondary school knows Descartes's vortices only through Bélise's simpering, and the state of astronomy in the seventeenth century through "that great long frightful spyglass" that Chrysale cannot stand.[13]

This indicates a lack of seriousness, a lack of truth in thought, which has no right to be presented as culture. Such evocations would have their place if they could be situated with respect to their real source, which would be grasped first, and not through the pharisaism of a work of art that has other objectives besides culture. The encyclopedic order of simultaneity is expelled from cultural education because it does not conform to the opinions of social groups, which never contain a representation of the order of simultaneity, because all they represent is a minimal fraction of life in a determined epoch, and they cannot situate themselves on their own. This hiatus between contemporary life and culture comes from the alienation

---

13. Molière, *The Learned Ladies*, Act II, scene 7, trans. Richard Wilbur (New York: Dramatists Play Service, 1977); Bélise and Chrysale are characters in this play. [TN]

of culture, which is to say from the fact that culture, in reality, is an initiation into the opinions of determinate social groups having existed in previous epochs; the primacy of literature in cultural education comes from this omnipotence of opinion; a work [*œuvre*], in particular a work that has survived, is in fact a work that has expressed the ethics of a group or of an epoch wherein this group could recognize itself; a literary culture is thus enslaved to these groups; it is at the level of these groups of the past. A literary work is a *social witness*. The entire share of didactic works is eliminated from culture, unless it is ancient, and can be considered as a witness to the didactic "genre." Contemporary culture feigns to consider the didactic genre as something extinct, when perhaps never before has there been as much expressive force, as much art, as much human presence in scientific and technical writings. It is really culture that has now become a genre with its fixed rules and norms; it has lost its sense of universality.

Therefore education, in order for it to be fully educational, lacks human dynamisms. Considering in particular the technical aspect of this education and of encyclopedism, it is plain to see that it constitutes a mediator of great value, since it has aspects that make it accessible to the child and others that adequately symbolize the successive stages of scientific knowledge; the stumbling block that cultural education runs up against when it wants to become encyclopedic is the difficulty attached to wanting to understand this science merely on the basis of discursive intellectual symbols. Technical realization, on the contrary, provides the scientific knowledge that serves as its principle of functioning, in the form of a dynamic intuition that can even be apprehended by a young child, and which is susceptible to becoming more and more elucidated, doubled by a discursive form of comprehension; truly discursive knowledge admits of no degrees, it is either immediately perfect or false because it is inadequate. Through technics, encyclopedism could thus find its place in the education of the child without requiring capacities for abstraction, which the young child does not fully have at its disposal. In this sense, the child's acquisition of technological knowledge can initiate an intuitive encyclopedism, grasped through the nature [*caractère*] of the technical object. The technical object in fact distinguishes itself from the scientific object because the scientific object is an analytical object, which aims at analyzing a unique effect in all its most precise conditions and characteristics, whereas the technical object, far from being situated in its entirety within the context of a particular science, is in fact at a point of concurrence of a multitude of data and scientific effects coming from the most varied domains, integrating what appear to be the most heteroclite forms of knowledge [*savoirs*], and which can in some cases not be coordinated intellectually, while they are indeed coordinated practically in the functioning of the technical object; it has been said that the technical object is the result of the art of compromise; what it is, is indeed an eminently synthetic structure, which cannot be understood in any other way than through the introduction of a synthetic schematism that presides over invention. The technical schema, which is a relation between several structures and a complex operation taking place through these structures, is by its very nature encyclopedic, since it leads to a circularity of knowledge, a synergy of elements of knowledge that are still theoretically heterogeneous.

Perhaps it should be noted that until the twentieth century, technics were incapable of assuming this relation between the encyclopedic work and the culture given to the child. For indeed it was almost impossible at the time to find, within technics, truly universal operations, including the schematisms of sensation or of thought. Today the existence of a technics of information gives technology an infinitely greater universality. Information theory places technology at the center of a large number of diverse sciences, such as physiology, logic, aesthetics, phonetic or grammatical study and even the semantics of languages, numerical calculus, 155 geometry, the theory of organization of groups and of regimes of authority, the calculus of probability, and all the technics of transport of spoken, acoustic or visual information. Information theory is an inter-scientific theory that enables the systematization of scientific concepts as much as the systematization of the schematisms of various technics; information theory mustn't be considered as a technics among technics; in reality it is a thinking that acts as mediator between the various technics on the one hand, between the various sciences on the other, and finally between the sciences and technics; it can play this role because there are relations between the sciences that are not only theoretical, but also instrumental and technical, each science being capable of making use of a certain number of other sciences for its own benefit, which it uses as technical sources in order to carry out the effect it studies; a technical relation takes place between the sciences; technics, moreover, can be theorized in a scientific form; information theory intervenes as a science of technics and as a technics of the sciences, determining a reciprocal state of these exchange functions.

It is at this level, and at this level only, that encyclopedism and technical education can meet, within a coherence of two simultaneous and successive orders of universality.

We can thus say that, if to this day technics could only provide two difficult-to-reconcile dynamisms — one of which is geared toward the adult and the other toward the child, with information theory — then this antagonism today gives way to a mediating discipline that establishes continuity between specialization and encyclopedism, between the education of the child and that of the adult. What is thereby founded is a reflexive technology above and beyond the different technics, and what is thereby defined is a thinking that creates a relation between 156 the sciences and technics.

The consequence of this reflexive unification of technics and the end of the opposition between theoretical knowledge and practical knowledge for the reflexive conception of man is considerable; once this level has been reached, there is effectively no longer a hiatus or an antagonism between the time of education and adulthood; the order of succession and the order of simultaneity organize themselves in a relation of reciprocity, and adulthood is no longer antagonistic with respect to that of education. To a certain extent even the evolution of societies, stuck, up till now, on a determinism of youth, then of maturity and finally of old age, along with the political and social regimes corresponding to each, can no longer be conceived as fatal if the penetration of technics is deep enough to introduce a system of references and values that are independent of this implicit biologism.

A careful analysis of the dualisms in value systems, such as those between the manual laborer and the intellectual, the peasant and the citizen, the child and the adult, would show that underneath these oppositions there is a technical reason for the incompatibility between several groups of schematisms; the manual laborer is the one who lives according to an intuitive schematism at the level of material things; the intellectual, on the contrary, is the one who has conceptualized the sensible qualities; he lives in accordance with an order that stabilizes the order of succession by way of definitions of nature and the destiny of man; he holds a certain power of conceptualizing and of valorizing or devaluing human gestures and values lived at the level of intuition. The manual laborer lives according to the order of simultaneity; he is an autodidact when he wants to gain access to a culture. It is according to this same difference between schematisms that the man of the countryside is opposed to the city dweller. The man of the countryside is contemporary with a set [*ensemble*] of requirements and participations that make him a 157 being who is integrated into a natural system of existence; his tendencies and his intuitions are the links of this integration. The city dweller is an individual being, linked to a social coming-into-being rather than to a natural order. He is the opposite of the man of the countryside in the way that an abstract and cultivated being is the opposite of an integrated and uncultivated being. The city dweller is of a time, whereas the man of the countryside is of a place or region; the former integrates himself into the order of succession, the latter, into the order of simultaneity. The attachment to traditions by the man of the countryside is generally what is noted; but tradition is precisely the most unconscious aspect of historicity, which masks the representations of the successive order, and which supposes an invariance of succession. Real traditionalism is based on the absence of a representation of the series of coming-into-being; this coming-into-being is repressed [*enfoui*]. In the end, the opposition between the child and the adult summarizes these antagonisms; the child is the being of succession, made of virtualities, modifying himself in time and being aware of this modification and of this change. The adult, capable of facing the simultaneity of problems that life confronts him with because of his education, integrates himself into society according to the order of simultaneity. However, this maturity can only be fully achieved to the extent that society is stable and is not evolving too rapidly, otherwise a society that is in the process of transforming itself, in privileging the order of succession, communicates a dynamism to its adult members that turns them into adolescents.

CHAPTER TWO

THE REGULATIVE FUNCTION OF CULTURE IN THE RELATION BETWEEN MAN AND THE WORLD OF TECHNICAL OBJECTS. CURRENT PROBLEMS

I. – The different modalities of the notion of progress   159

The encyclopedists' attitude toward technics can be considered an enthusiasm roused by the discovery of the technicity of the elements. Machines are not, in fact, directly considered automata by the encyclopedists; rather, they are considered an assemblage of elementary devices. Diderot's collaborators directed their attention essentially to the organs of machines. In the eighteenth century, the technical ensemble was still at the scale of the cork cutter's workshop or that of the scale maker's; this ensemble links up with technical elements through the intermediary of the craftsman who uses tools or machine-tools, rather than through the intermediary of veritable technical individuals. The division of subjects for study is consequently made according to rubrics of utilization and not according to schemas of technics, i.e., according to types of machines; the principle in grouping and analyzing technical beings is the denomination of the trade, rather than that of the machine. Very different trades, however, can make use of identical or almost identical tools. This principle of grouping thus leads to a certain superfluity of the presentation of tools and instruments which, from one illustration to the next, can be closely related forms.

The principle of grouping according to technical ensembles comprised of an indefinite plurality of elements, however, is linked very closely to the idea of *continuous progress* such as it existed for the Encyclopedists. It is when technicity is grasped at the level of elements that technical evolution can occur according to a continuous line. There is a correlation between a molecular mode of the existence of technicity and a continuous pace of the evolution of technical objects. Gears and screw threads were cut better in the eighteenth century than in the seventeenth century; from this comparison between the same elements made in the seventeenth and in the eighteenth century arose the idea of the continuity of progress as a forward march in what we have called the concretization of technical objects. The evolution of this element, which takes place within already constituted technical ensembles, does not provoke any upheaval: it improves the results of fabrication without brutality, and authorizes the craftsman to preserve habitual methods, while experiencing the feeling of facilitation at work; the habitual gestures, better served by more precise instruments, now yield better results. The optimism of the eighteenth century is to a large extent based on the elementary and continuous improvement of the conditions of technical work. Anxiety effectively arises from those transformations that provoke a break within the rhythms of everyday life, making the old habitual gestures useless. But the improvement of the tool's technicity plays a euphoric role. When man, while preserving the fruit of his training, exchanges an old tool for a new tool whose manipulation is the same, he has the feeling of having more precise, skillful, and rapid gestures; it is the entire corporeal schema that expands against his limitations, that dilates and frees itself; the impression of awkwardness diminishes: the trained man feels more skillful with a better tool; he has greater self-confidence; for the tool is an extension of the organ, and is carried by the gesture.

161

The eighteenth century was the pivotal moment for the development of tools and instruments, if by *tool* one understands the technical object enabling one to prolong and arm the body in order to accomplish a gesture, and by *instrument* the technical object that enables one to prolong and adapt the body in order to achieve better perception; the instrument is a tool of perception. Some technical objects are both tools and instruments, but they can be called tools or instruments according to the predominance of their active function or of their perceptive function: a hammer is a tool, even though, through the receptors of kinaesthetic and vibratory tactile sensitivity, we can subtly perceive the instant when the nail starts to writhe or to split the wood and penetrate it too fast; the hammer must effectively act on the tip so as to drive it in, so that, according to the manner in which this operation of driving in the tip is executed, definite information is communicated to the senses of the one who holds the hammer in his hand; the hammer is thus first a tool, since it is as a result of its tool-function that it can serve as an instrument; even when the hammer is used as a pure instrument, it is still, primarily, a tool: the mason recognizes the quality of a stone with his hammer, but for this to happen the hammer must first partially chip away at the stone. Conversely, a telescope or a microscope are instruments, in the same manner as a level or a sextant are: these objects serve to collect information without accomplishing any prior action on the world. And the eighteenth century is the age in which both tools and instruments were made with greater care, reaping the rewards of seventeenth century discoveries within static and dynamic mechanics, as well as those found in geometrical and physical optics. The undeniable progress of the sciences was translated into the progress of technical elements. This accord between scientific investigation and technical consequences is a new reason for optimism that adds itself to the content of the notion of progress, through the spectacle of this synergy and this fecundity of the domains of human activity: the instruments, improved by the sciences, are at the service of scientific investigation. 162

The aspect of technical evolution changes, on the contrary, when the birth of complete technical individuals is encountered in the nineteenth century. As long as these individuals merely replace animals, the perturbation is not a frustration. The steam engine replaces the horse in hauling wagons; it drives the spinning mill: gestures are modified to a certain extent, but man is not replaced insofar as the machine simply provides a greater utilization of energy sources. The Encyclopedists were familiar with the windmill, which they magnified and represented as dominating the landscape from the height of its tall silent structure. Several extremely detailed illustrations are dedicated to new and improved water mills. Man's frustration starts with the machine that replaces man, with the automatic weaving loom, with the forging press, with the equipment of the new factories; what the worker destroys during a riot are the machines, because they are his rivals; the machine is no longer simply an engine but a bearer of tools; eighteenth century progress left the human being intact because the human individual remained a technical individual among his tools of which he was both the center and bearer. It is not necessarily through its size that the factory distinguishes itself from the craftsman's workshop, but through the change in relation between the technical object and the human being: the factory is a technical ensemble that is comprised of automatic machines, whose activity is parallel to that of human activity; the factory uses true technical individuals, whereas, in the workshop, it is man who lends his individuality to the accomplishment of technical actions. From then on the most positive, most direct aspect, of the first notion of progress, is no longer 163 experienced [*éprouvé*]. The progress of the eighteenth century is a progress experienced by an individual through the force, speed, and precision of his gestures. The progress of the nineteenth century can no longer be experienced by the individual, because it is no longer centralized with the individual as the center of command and perception in the adapted action. The individual becomes the mere spectator of the results of the functioning of the machines, or the one who is responsible for the organization of technical ensembles putting the machines to work. This is why the notion of progress splits in two, becomes aggressive, ambivalent, and a source of anxiety; progress is at a remove from man and no longer makes sense for the individual, because the conditions of the individual's intuitive perception of progress no longer exist; this implicit judgment, which is very close to that of kinesthetic impressions and to the facilitation of a corporeal dynamism which formed the basis of the notion of progress in the eighteenth century, disappears, except within domains of activity in which the progress of the sciences and of technics provides, as in the eighteenth century, an extension and facilitation of individual conditions of action and observation (as is the case with medicine and surgery).

Progress is henceforth thought of as cosmic, at the level of its overall results. It is thought abstractly, intellectually, in a doctrinal manner. Progress is no longer thought by craftsmen, but by mathematicians, who conceive of progress as man taking possession of nature. Beginning with the Saint-Simonians, this idea of progress starts to support technocratism. An idea of progress that was conceived and desired substitutes itself for the impression of progress as something undergone [*éprouvé*]. The individual who thinks progress is not the same individual as the one who works, except in some rather rare cases, such as the case of the printer and lithographer, who have mostly remained craftsmen. Even in these cases, for those who think deeply about its nature, the advent of the machine is expressive of an aspiration for the transformation of social structures. One could say that work and technicity were linked in the eighteenth century through the experience [*épreuve*] of elementary progress. Conversely, the twentieth century brings about the disjunction of the conditions for the intellection of progress and for the experience of the internal rhythms of work resulting from this same progress. Nineteenth century man does not experience progress as a worker: he experiences it as an engineer or a user. In fact, the *engineer*,[14] the man of the machine, becomes the organizer of the ensemble made up of workers and machines. Progress is grasped as a movement that manifests itself through its results, rather than as progress in itself understood as the ensemble of operations that constitute it, as the elements that actualize it, and as being valid for a large number of people that would be coextensive with humanity.

Indeed the poets of the end of the first half of the nineteenth century keenly felt

14. English in original. [TN]

progress to be the general march of humanity, with its charge of risk and anxiety. Within this progress there is something of an immense collective adventure, of a voyage and even of a migration toward another world. This progress contains at once something triumphant and crepuscular. It is perhaps the word that Vigny sees written above the cities in *La Maison du berger* [The Shepherd's house]. The feeling of ambivalence toward the machine can be found in the evocation of the locomotive and the compass, the former in *La Maison du berger*, the latter in *La Bouteille à la mer* [The Bottle in the sea]. The latter poem shows how Vigny felt about the transient (and perhaps transitory because contradictory) nature of progress in the nineteenth century. This unfinished, incomplete idea of progress, contains a message for posterity; it cannot fulfill itself within itself. One of the aspects of *Les Destinées* [Destinies] is to accept living within this moment of technical evolution. Vigny made it accurate and significant by understanding that technical evolution could not satisfy itself by way of itself, that it couldn't simply close in on itself.

A third aspect of the notion of technical progress emerges with the repercussions of the internal regulation of technical individuals regarding technical ensembles, and, through these, regarding humanity. The second stage, that which corresponds 165 to the arrival of a new wave of technics at the level of individuals, was characterized by the ambivalence of progress, by dual situation of man with regard to the machine, and by the production of alienation. This alienation grasped by Marxism as having its root in the relation of the worker with the means of production, does not only derive, in our view, from a relation of property or non-property between worker and the instruments of work. Beneath this juridical and economic relation exists an even more profound relation, that of the continuity between the human individual and the technical individual, or of the discontinuity between these two beings. The reason why alienation arises is not solely because in the nineteenth century the human individual who works is no longer the owner of his means of production, whereas in the eighteenth century the craftsman was the owner of his instruments of production and of his tools. Alienation does indeed emerge the moment the worker is no longer the owner of his means of production, but it does not emerge solely because of this rupture in the link of property. It also emerges outside of all collective relation to the means of production, at the physiological and psychological level of the individual properly speaking. The alienation of man in relation to the machine does not only have a socio-economic sense; it also has a physio-psychological sense; the machine no longer prolongs the corporeal schema, neither for workers, nor for those who possess the machines. Bankers whose social role has been exalted by mathematicians such as the Saint-Simonians and Auguste

Comte are as alienated in their relation to the machine as the members of the proletariat. What we mean by this is that there is no need to presuppose a master-slave dialectic in order to account for the existence of alienation within the proprietor class. The relation of property with respect to the machine contains as much alienation as the relation of non-property, even if it corresponds to a very different social state. On either side of the machine, above and below, the worker, who is a man of elements, and the industrial boss, who is a man of ensembles, both lack a true relation with the individualized technical object in the form of the machine. Labor and capital are two modes of being where one is as incomplete as the other with respect to the technical object and the technicity contained in industrial organization. Their apparent symmetry does not at all mean that the union of capital and of labor reduces alienation. The alienation of capital is not alienation with respect to labor, with respect to the contact with the world (as in the master/slave dialectic), but rather with respect to the technical object; the same goes for labor; what labor lacks is not what capital possesses, and what capital lacks is not what labor possesses. Labor possesses the intelligence of elements, capital possesses the intelligence of ensembles; but it is not by combining the intelligence of elements with the intelligence of ensembles that one can arrive at the intelligence of the *intermediary and non mixed* being that is the technical individual. Element, individual, and ensemble follow each other along a temporal line; the man of the element is late with respect to the individual; but the man of ensembles who has not understood the individual is in no way ahead of his time with respect to it; he tries to enclose the present technical individual in the structure of an ensemble belonging to the past. Labor and capital are both late with respect to the technical individual that is a depository of technicity. The technical individual is not of the same era as the labor that enacts it and the capital that frames it.

The dialogue between capital and labor is false because it is of the past. The collectivization of the means of production cannot achieve a reduction of alienation on its own; it can only achieve this reduction if it is the precondition for the acquisition of the intelligence of the individuated technical object by the human individual. This relation between the human individual and the technical individual is the most difficult to form. It presupposes a technical culture, which introduces the capacity of different attitudes rather than that of work and of action (work corresponding to the intelligence of the elements and action to the intelligence of ensembles). What work and action have in common is the predominance of finality over causality; in both cases, the effort is directed at a certain result to be obtained; the employment of means finds itself in the position of minority with respect to the result: the schema of action matters less than the result of the action. In the technical individual, however, this disequilibrium between causality and finality disappears; viewed from the outside the machine is *made in order to* obtain a certain result; but, the more the technical being becomes individualized, the more this external finality effaces itself for the benefit of the internal coherence of functioning; the functioning is finalized with respect to itself before being so in its relation with the external world. Such is the automatism of the machine, and such is its self-regulation: there is, at the level of regulations, a functioning, and not only a causality or finality; in self-regulated functioning, all causality has a sense of finality, and all finality a sense of causality.

II. – Critique of the relation between man and the technical object
as it is presented by the notion of progress arising from thermodynamics
and energetics. Recourse to Information Theory

Within man, what can be in a relation with the individualized technical being is the intuition of the schemas of functioning; as a being who participates in its regulation, man can be coupled to the machine as an equal, and not merely as a being who directs or utilizes it through the incorporation of ensembles, or as a being who serves it by supplying matter and elements. By this we mean to say that neither an economic theory nor an energetic theory can account for such a coupling of man   168 and machine. Economic or energetic links are too external for it to be possible to define this true coupling through them. There is an inter-individual coupling between man and machine when the same self-regulating functions are better and more subtly accomplished by the man-machine couple than by man or machine alone.

Let's take the case of what we call memory. Leaving aside all the mythological assimilations of vital functions to artificial operations, one can say that man and machine present two complementary aspects of the use that is made of the past. The machine is capable of retaining monomorphic documents that are very complex, richly detailed, and precise for a very long time. A magnetic tape three hundred meters long can keep a recording of the magnetic translation of any noise or sound in a bandwidth of 50 to 10,000 Hz, corresponding to about an hour's listening-time, or two hours if one accepts a reduction of the frequency band to an upper limit of 5,000 Hz. A roll of film of the same size can record scenes that take about half an hour to play, with a definition of about 500 lines, that is to say in a way that makes it possible to distinguish about 250,000 pixels on every image. Magnetic tape can thus record 3,600,000 sonic events, each one distinct from the others; cinematographic film stock can record 120 million pixels, each one distinct from the others. (The difference in these figures does not derive solely from the fact that the grain of a magnetic strip is larger than that of the sensitive film; in fact, it is of the same size; it comes primarily from the fact that the recording of sound corresponds to a linear track on a tape, whereas the recording of images corresponds to the splicing of successive surfaces in which almost every one of the sensitive pixels can become a carrier of information.) Now, what characterizes the conservation function of this machine is that it is absolutely without structure; film does not record clear-cut figures, geometrical images for instance, any better than the disordered image of grains in a pile of sand; to a certain extent the vivid oppositions of clear cut surfaces are even less well recorded than the disordered uniformity of the grains of sand, because of the phenomena of light diffusion in the film base, which create the so-called halo effect, around very bright and clearly delineated areas. Similarly, magnetic tape does not record continuous and well-formed musical sounds any better than transitional tones or noises: there is no order in the preservation of recordings by the machine, which does not possess the faculty of selecting forms. Human perception distinguishes forms, perceptual units, when looking at or listening to recorded documents. But the recording itself does not really contain these forms. The inability of the data-preserving function of the machines is relative to the recording and reproduction of forms. This incapacity is general, it exists at every level. A considerable intricacy is needed in order for a calculating machine to be capable of writing results in clearly legible figures on a cathode ray tube screen. The *numeroscope* is made of very delicate and complex combinations, using encodings in order to obtain lines that can somehow reproduce numbers. It is far easier to reproduce Lissajous figures than to write the number 5. The machine cannot retain forms, but merely a translation of forms, by means of an encoding in a spatial or temporal distribution. This distribution may be very durable, like that of a magnetic tape, definitive, like that of silver grains in chemically sensitive film, or altogether provisional, like that of the series of pulses passing through a mercury column with a piezo-electric crystal at each end, used in certain types of calculating machines for the preservation of partial results during the course of the operation; it can also be very fleeting but sustained, as in the case of the recording of numbers on a mosaic in a certain type of cathode ray tube somewhat like the iconoscope, and equipped with two electron guns, one for reading and inscription, the other for upkeep (the Massachusetts Institute of Technology RCA Selectron and storage tube). The plasticity of the support must not be confused with a true plasticity in the recording function; it is possible to erase the numbers inscribed on the beryllium mosaic of the selectron to the thousandth of a second, and to replace them with others, but the speed with which successive recordings succeed each other on this same support in no way means that the recording itself is plastic; each recording, taken in itself, is perfectly rigid. It is obviously possible to erase the magnetization of the oxide grains from the magnetic strip in order to make another recording. But the new recording is completely separate from the preceding one; if the first is poorly erased, it interferes with the next recording, blurring it, rather than facilitating it.

In human memory, on the contrary, it is the form that is preserved: preservation itself is only a limited aspect of memory, which is the power of selecting forms, of schematizing experience. The machine would be capable of performing a similar function only if the already-recorded magnetic tape was superior to a new tape in recording certain sound figures, which is not the case. The plasticity in the memory of machines is that of the medium, whereas in human memory it is the plasticity of the content itself.[15] One can say that, in man, the function of preserving recollections resides in memory, because memory, conceived as a set [*ensemble*] of 171 forms, of schemas, receives the recollection it records because memory connects the recollection with its forms; in a machine, on the contrary, recording occurs without prior memory. From this essential difference comes human memory's significant incapacity for fixing elements without order. It would take a very long time to learn the relative positions of fifty tokens of different colors and shapes emptied on a table in disorder; even a blurred photographic view is more useful than a human witness when it comes to affirming the relative position of various objects in space. Machine memory triumphs in multiplicity and disorder; human memory triumphs in the unity of forms and in order. Every time a function of integration or comparison appears, the most complex and best-built machine provides results that are considerably inferior to those that human memory can achieve. A calculating machine can be coded in order to translate, but its translation remains very elementary and crude. It presupposes a prior reduction of each of the two languages on a simplified basis, along with a reduced vocabulary and a fixed number of turns of phrase. For the machine lacks the plasticity of integration, which is the vital aspect of memory through which it distinguishes itself in that very instant

---

15. An unused magnetic tape is equal to or superior to an already used tape, even if only for registering the same form several times in a row. A cathode ray tube on which the same image is continually fixed, far from becoming better at registering it, loses its sensitivity in areas occupied by the image, so much so that after prolonged use it is more sensitive to new images, which do not form at the same points as the older ones.

from machine memory: the *storage*[16] of the calculator or the translating machine (which is nothing but a classical calculating machine coded in a certain way) is very

172   different from the function of the present through which memory exists, in man, at the very level of perception, through perception, making sense of the present word according to the general turn of phrase and of earlier phrases, or even according to the totality of experience acquired in the past regarding the person who speaks. Human memory receives contents that have a formative power in the sense that they overlap with themselves, are grouped, as if acquired experience served as a code for new acquisitions, in order to interpret and fix them: in man and more generally in the living being *content becomes coding*, whereas in the machine coding and content remain separate as condition and conditioned. Content introduced into human memory will superimpose itself on prior content and take form on it: the living is that in which the *a posteriori* becomes *a priori*; memory is the function by which *a posteriori* matters become *a priori*.

A complex technical operation, however, requires the use of both forms of memory. Non-living memory, the memory of the machine, is useful in cases where faithfulness in the preservation of details outweighs the syncretic nature of a recollection integrated into experience, having signification through its relation with other elements. Machine memory is the memory of the document, of the result of measurement. Man's memory is that which, after an interval of many years, evokes a situation because it involves the same significations, the same feelings, the same dangers as another, or simply because this similarity makes sense according to the implicit vital coding constituted by experience. In both cases memory allows self-regulation; but human memory enables a self-regulation according to an ensemble of significations valid in the living and capable of developing only in the latter; that of the machine grounds a self-regulation that makes sense in the world of non-living beings. The significations according to which human memory

173   functions stop where those according to which machine memory functions begin.

The coupling of man to machine begins to exist from the very moment when a coding common to both of these memories can be discovered, in order for a partial convertibility of one into the other to be realized, so that a synergy can become possible. One case of this coupling is provided by the permanent data file of telephone calls. The summarized information recording up-to-date results recently obtained from multiple domains classed under different rubrics is recorded on magnetic tapes. A catalog and a telephone call system make it possible, through the use of selectors, to rapidly obtain a read-out of what has been recorded on any

---

16. English in original. [TN]

of the magnetic tapes. Here, human memory is where the words and names of the rubrics have signification. The machine, on the contrary, is where a series of definite pulses provokes one magnetic tape player to be powered and not another: this fixed and rigid faculty for selection is very different from the one that prompts the inquirer to dial one particular phone number rather than another. This pure case of coupling between machine and man helps us to understand the mode of coupling that exists in other cases: there is coupling when a single and complete function is carried out by both beings. Such a possibility exists every time that a technical function has a defined self-regulation. Functions that contain a self-regulation are the ones where the accomplishment of a task is directed not only by a model to be copied (according to an end), but by a partial result of the accomplishment of the task, intervening as a condition. In the artisanal operation, this control through information gathering is frequent; man being at once the mover [*moteur*] tool and the perceiving subject regulates his action according to instantaneous partial results. The tool is at once tool and instrument, which is to say a means of action 174 prolonging the organs and a channel of recurring information. The machine, on the contrary, as a complete closed individual replacing man, generally has no system of self-regulation: it goes through the motions of a stereotypy of successive gestures according to a predetermined conditioning. This first type of machine is what one could call a mechanical being without self-regulation. It is indeed a practical technical unit, but not a technical individual strictly speaking.

Despite appearances, it is, on the contrary, the truly automatic machine that least replaces man, because the function of regulation that exists in this machine presupposes a variability in its operation, an adaptability of its functioning to the completion of this work. A rather elementary enthusiasm for self-regulating automata lets us forget that these machines are precisely the ones that are most reliant on man; while the other machines need man simply as a servant or organizer, self-regulating machines need man as a technician, which is to say as an associate; their relation to man is situated at the level of this regulation, not at the level of elements or ensembles. But it is through this regulation that automatic machines can be linked to the technical ensemble in which they function. Just as the human individual is not linked to the group by his elementary functions, whether active or perceptive, but by his self-regulation which gives him his personality and character, so the machine is integrated with the ensemble not only in an abstract and preliminary way, by its function, but also, at every moment, by the way it performs its own task according to the requirements of the ensemble. There is no such thing as a purely internal, entirely isolated self-regulation; the results of the action are not merely results in themselves but also in relation to their external milieu, to the ensemble. Now this aspect of self-regulation in which *the whole of the milieu must be taken into account* cannot be achieved by the machine alone, even if it is perfectly automated. The type of memory and the type of perception needed for this aspect of regulation call for the integration and transformation of the *a posteriori* into the *a priori* which only the living thing can achieve within itself. There is something alive in a technical ensemble, and the integrative function of life can be ensured only by human beings; the human being has the capacity to understand the functioning of the machine, on the one hand, and the capacity to live, on the other: one can speak of technical life as being that which actualizes this relation between these two functions in man. Man is capable of taken upon himself the relation between the living being that he is and the machine he fabricates; the technical operation requires both technical and natural life.

Technical life, however, does not consist in overseeing machines, but in existing at the same level as a being that takes charge of the relation between them, capable of being coupled, simultaneously or successively, with several machines. Each machine can be compared to a monad, isolated in itself. The capacities of the machine are only those that have been put in it by its constructor: it unfolds its properties just as substance develops its attributes. The machine results from its essence. Man, on the contrary, is not a monad, for in him the *a posteriori* becomes *a priori*, the event becomes a principle. Man as technician does not perform this function prior to the manufacture of machines, but during their operation. He fulfills the function of the present, and maintains the correlation because his life consists of the rhythm of the machines that surround him and that he connects to one another. He fulfills the function of integration, and prolongs self-regulation beyond each monad of automatism through the interconnection and inter-commutation of monads. The technician is indeed in a certain sense the man of ensembles, but in a very different way from the one that characterizes the industrialist. The industrialist, in the same way as the worker, is pushed by finality: he targets a result; herein lies their alienation; the technician is the man of the operation in the course of its accomplishment; he does not take charge of directing the ensemble but rather guides its self-regulation during functioning. He absorbs within himself the sense of the work and the sense of the industrial direction. He is the man who knows the internal schemas of functioning and organizes them in relation to each other. On the contrary, machines are ignorant of general solutions and cannot resolve general problems. Whenever it is possible to replace a complex operation by a greater number of simple operations, this procedure is used in the machine; this is the case for calculating machines that use a binary system of numeration (rather than a decimal system) and reduce all operations to a series of additions.[17]

In this sense one can affirm that the birth of a technical philosophy at the level of ensembles is possible only through an in-depth study of regulations, which is to say of information. True technical ensembles are not the ones that use technical individuals, but those that form a fabric of technical individuals through a relation of interconnection. Any philosophy of technics that starts from the reality of ensembles using technical individuals without putting them into a relation of information remains a philosophy of human power through technics, not a philosophy of technics. One could use the term "autocratic philosophy of technics" for a philosophy that takes the technical ensemble as a place where machines are used in order to obtain power [*puissance*]. The machine is only a means; the end is the conquest of nature, the domestication of natural forces by means of a first act of enslavement: the machine is a slave whose purpose is to make other slaves. Such a dominating and enslaving inspiration can coincide with the quest for man's   177 freedom. But it is difficult to free oneself by transferring slavery onto other beings, men, animals, or machines; to reign over a people of machines that enslave the entire world is still to reign, and every reign presupposes the acceptance of the schemas of enslavement.

Technocratic philosophy itself is affected by an enslaving violence, insofar as it is technocratic. A technicism that comes from a reflection on autocratic technical ensembles is inspired by the unbridled will to conquer. It is immoderate, and lacks internal control and self-mastery. It is a force that is unleashed and which can only perpetuate itself for the duration of the ascending phase of success or conquest. Saint-Simonism triumphed under the Second Empire because there were train platforms to be built, railway track to be laid, bridges and viaducts to be spread across the valleys, and tunnels to be pierced through mountains. This aggressive conquest has the characteristics of a rape of nature. Man enters into possession of the earth's bowels, traverses and labors, trespasses onto what had previously been unreachable. Thus technocracy in a certain sense implies a violation of the sacred. To throw a bridge over an inland sea, to attach an island to the continent, to pierce an isthmus, is to modify the configuration of the earth, to undermine its natural integrity. There is a pride of domination in this violence, and man entitles himself as creator or at least foreman of creation: he plays the role of the demiurge. It is Faust's dream, being played out by an entire society, by all technicians.

---

17. The fundamental vital processes are, on the contrary, processes of integration.

Indeed, the development of technics is not enough for technocratism to emerge. Technocratism represents a will to acquire power that comes to light in a group of men possessing knowledge but not power, the knowledge of technics but not the money to put them to work or the legislative power to free themselves of all constraint. In France, technocrats are essentially poly-technicians,[18] which is to say men who, with respect to technics, find themselves in a situation of being intelligent users and organizers of technics rather than veritable technicians. These mathematicians think in sets [*ensembles*], not in individual operational units; what holds their attention is not so much the machine as enterprise.

178

Furthermore, and essentially, in an even more profound way, the conditioning arising from the state of technics comes to be added onto that of psychosocial conditioning. The nineteenth century could produce only a technological technocratic philosophy because it discovered engines and not regulations. It is the age of thermodynamics. However, in a certain sense, an engine is indeed a technical individual since it cannot function without a certain number of regulations, or at the very least a certain number of automatisms (intake, exhaust); but these automatisms are auxiliary; their function is to enable the renewal of the cycle. Sometimes the addition to stationary machines of veritable self-regulators, such as Watt's governor (a centrifugal regulator, called a ball regulator), individualizes the heat engine in a very complete way; regulators nevertheless remain accessories. When a thermal machine has to generate great momentum, in accordance with an extremely intermittent working condition [*régime*], it is good to have a man keeping watch nearby who can press the regulator-lever before the load increases, because the regulator, by intervening with too much of a delay, risks intervening when the engine has already slowed down because of the sudden load increase: this is what happens when a steam engine is used to cut large tree trunks into planks; once the regulator kicks in without human intervention, the circular saw has already stalled, or the belt has already fallen: the operator hits the regulator-lever a half-second before the saw-wheel cuts into the trunk; the engine thus functions at full speed and is accelerating when the work load suddenly increases. On the other hand, the Watt regulator is extremely efficient and precise when load variations become slower and gradual. Such incapacity for rapid variations can be explained by the fact that, in thermodynamic engines, even when there is self-regulation, it has no information channels that are separate from the effectors. There is effectively a feed-back information channel in the Watt governor, a channel of counter-reaction (*feed*-back),

179

18. An allusion to France's elite École polytechnique (whose students and alumni are called *polytechniciens*) and to other polytechnic schools founded, along with other *Grandes Écoles*, in the long aftermath of the Revolution in order to train the new meritocratic elite. [TN]

but this information is transmitted through a channel that is not distinct from the one used by the motive power that allows the engine to move a resistant organ: the regulator is connected to the output shaft; the whole ensemble made up of the fly-wheels, the main shaft, the volumetric cylinder device and then the system for transforming alternating movement into circular movement must therefore already have slowed down through the loss of its kinetic energy so that the regulator can intervene by increasing the admission time of the engine and consequently also its power. Now, there is a serious drawback in this lack of distinction between the motive power channel (the energy channel) and the feedback channel (the information channel), which greatly reduces the effectiveness of regulation, and the extent of the individualization of the technical being: when the engine slows down (which is necessary in order for the regulator to come into action), the decrease in running speed causes a decrease in power (the engine power, in low or medium working conditions [*régime*], where the throttling of steam in the sliding valve does not come into play, is proportional to the sum of all the elementary work performed by successive piston strokes in a unit of time). The decrease in angular speed results in the deterioration of the very conditions of reactivity that the regulator is designed to provoke.

It is this lack of distinction between the energy channel and the information channel that marks the thermodynamic age, and constitutes the limit of the individualization of thermal engines. Suppose on the contrary that a gauge measures the torque on the transmission shaft at the outlet of a heat engine at every moment, 180 and that the result of this measurement is sent back to steam intake (or fuel intake or carburized air-intake if it is an internal combustion engine), so as to increase steam-intake in relation to the increased resistance exerted on the transmission shaft; the pathway by which the resistance measurement then is fed back into the steam supply and modifies it is thus distinct from the energy channel (steam, cylinder, piston-rod, crank shaft, axle, transmission shaft); there is no need for the engine to slow down for its power to increase: the information sampling rate along the information channel can be extremely high with respect to the time constants of the energy channel, for instance one every few hundredths or a few thousandths of a second, while the cycle of a stationary steam engine lasts about a quarter of a second.

It is therefore natural that in machines, the advent of the use of information-channels that are distinct from energy channels caused a very profound change in the philosophy of technics. This advent was conditioned by the development of the vehicles of information, and in particular of low current signals. This is what we call electric currents considered not as energy carriers, but as vehicles for information. Electric current, as a vehicle for information, is equal only to Hertz waves or a light beam, which also consists of electromagnetic waves like Hertz waves: because electric current and electromagnetic waves have in common both an extreme speed of transmission and the ability to be modulated with precision, without noticeable inertia, in frequency as well as in amplitude. Their ability to be modulated makes them faithful carriers of information, and their speed of transmission makes them rapid carriers. What then becomes important is no longer the power conveyed, but the accuracy and fidelity of the modulation transmitted by the information channel. Beyond the dimensions defined by thermodynamics, a new category of physical dimensions emerges that makes it possible to classify information channels and compare them. This elaboration of new concepts has a particular sense for philosophical thought because it provides the example of new values which, until this day, made no sense in technics, though they made sense in human thought and behavior. Thermodynamics had thus defined the notion of efficiency for a conversion system such as an engine: the efficiency is the ratio between the amount of energy at the inlet of the engine and the amount of energy collected at the outlet; between the inlet and the outlet there is a change in the form of energy; for example, in the case of the heat engine, thermal energy becomes mechanical energy; since we know the mechanical equivalent of a calorie, we can define the efficiency of an engine as a transformer of thermal energy into mechanical energy. In every device that performs a conversion, we can more generally define an efficiency that is the ratio between two energies; there is thus the efficiency of the furnace, which is the ratio between the amount of chemical energy contained in the fuel system and the amount of heat actually released; the efficiency of the furnace-boiler system, defined by the ratio between the caloric energy produced by the furnace and the thermal energy effectively transmitted to the water in the boiler; there is an engine efficiency that is the ratio between the energy contained in the system that is constituted by the hot steam sent to the inlet and by the cold sink in the condenser, and the mechanical energy effectively produced by the steam expansion in the cylinder (a theoretical efficiency governed by the Carnot principle). In a series of energy transformations, the efficiency calculated between the first energy input and the last energy output is the product of all partial efficiencies. This principle applies even where the energy collected in the output is of the same nature as that at the inlet; when a storage battery is being charged, there is an initial partial efficiency which is the conversion of electrical energy into chemical energy; when discharging it, there is a second partial efficiency, which is that of the conversion of chemical energy into electrical energy: the output of the battery is the product of these two efficiencies. However, when an information channel is used to transmit information, or when information is recorded onto a medium for storage purposes, or when there is a transfer from one information medium to another medium (for example from mechanical vibrations to an alternating current in which the frequencies and amplitudes follow the vibrations), a loss of information occurs: what is collected at the outlet is not identical to what was at the inlet.

For example, if one wants to transmit a current of acoustic frequencies through an information channel that is a telephonic circuit, one notes that some frequencies are correctly transmitted: for these, the modulation collected at the outlet is identical to that which was put in at the input of the circuit. But the bandwidth of the telephone circuit is narrow; if one enters a noise or a complex sound at the entrance to this channel a considerable deformation ensues: the modulation collected at the outlet is not at all comparable to that which was entered at the input; it consists in an impoverishment of the former; for example, the fundamental frequencies of complex sounds between 200 Hz and 2000 Hz are transmitted correctly, but deprived of their upper harmonics. Or once again, the circuit introduces a harmonic distortion, which is to say that a sinusoidal sound that was entered at the input is no longer represented by a sinusoidal tension at the outlet; despite their apparent difference, by the way, the two phenomena are the same: a circuit that introduces a harmonic distortion is an information channel with a narrow   183 bandwidth that would transmit, without appreciable distortion, a sound that has the same frequency at the input as that of the harmonic component which appears at the outlet even though it wasn't at the input, which is something that can occur when the circuit has a resonance at this harmonic frequency. A perfect information channel would provide all the modulations, however rich or complex they may be at the outlet, as had been put in at the input. One could attribute to this information channel an efficiency equal to 1, as one would attribute to a perfect engine.

These efficiency characteristics of information channels are not energy characteristics, and very often good information efficiency goes hand-in-hand with low energy efficiency: an electromagnetic loudspeaker has better energy efficiency than an electro-dynamic loudspeaker, but very poor information output. This fact is fairly well explained if one considers that, in a transformation system, the best energy efficiency is obtained when there is a tight coupling of two elements by a sharp resonance; a transformer whose coil capacitances are in tune with a certain frequency has an excellent primary-secondary coupling for this frequency; but it has a poor coupling for the other frequencies: it therefore transmits this frequency selectively, which causes a considerable signal loss when it is being used for broadband transmission; the energy output of a transformer designed to transmit information is lower, but constant for a wide frequency band. Energy efficiency and information output are thus two dimensions that are not linked to each other: the technician is often obliged to sacrifice one of the two outputs to obtain the other. Form is what is essential in information channels, and the conditions of its correct transmission are very different than those of high efficiency energy transmission. The resolution of problems related to information channels calls for an attitude that is different in spirit to that which is appropriate for problem solving in applied thermodynamics.[19] A technician of thermodynamics tends toward gigantism in constructions and large-scale effects, because thermodynamic efficiency increases with the size of engines and installations. It is certainly possible to construct a small-scale steam engine, but its efficiency is low; even if it is very well constructed, it cannot attain an excellent efficiency because heat loss and the importance of mechanical friction significantly come into play. The turbine is a system for the transformation of thermal energy into mechanical energy, which offers a better efficiency than that of a reciprocating engine; but if a turbine is to function properly, it requires a large installation. The efficiency of three small thermal power plants is lower than that of a single power plant with the same power as the three small ones together. This increase in efficiency with the size of the involved machines is a general and practical law of energetics that exceeds the framework of thermodynamics strictly speaking; an industrial electric transformer generally has a higher efficiency than one with a nominal power of 50 watts. However, this tendency is much less pronounced within new forms of energy, such as electrical energy, than within the old ones, such as heat; nothing would stand in the way of constructing a small scale high-output electric transformer; if the efficiency of low power devices is somewhat neglected, then this is because the loss of output is less important for them than for industrial devices (heat, in particular, is more easily dissipated, for the same reasons that a small steam engine has a lower efficiency than a large one).

On the contrary, the information technician is brought in to find the smallest possible dimensions compatible with the residual thermodynamic requirements of the devices he uses. Information, effectively, is all the more useful, in a regulation process, in that it intervenes without delay, since the increase in size of information machines or transmission devices increases inertia and transit time. The stylus of the telegraph has become too heavy; a cable can transmit far more signals than the stylus can print; a single cable could carry the traffic of thirty simultaneous calls.

---

19. Or, more generally, in energetics.

In an electronic tube, the time of transit for electrons between cathode and anode is responsible for a cutoff of high frequencies; the smallest electronic tube is the one that can reach the highest frequencies, but this same tube then has a very low power, because its small size does not allow it to evacuate enough heat without reaching a temperature that would compromise its functioning. It is possible that one of the causes of the tendency toward size reduction, which we have witnessed since 1946, lies in the discovery of this imperative of information technics: to build technical individuals and above all elements of a very small size, because they are more perfect and have better information output.

III. – Limits of the technological notion of information in order
to account for the relation between man and the technical object.
The margin of indeterminacy in technical individuals. Automatism.

A philosophy of technics, however, cannot be founded exclusively on an unconditional quest for form and efficiency of form in the transmission of information. Yet the two kinds of efficiency, which appear to diverge, and which in fact already 186 diverge at the very beginning, are once again encountered further on: when the quantity of energy that serves as a carrier for information tends toward a very low level, a new type of output loss appears, one that is due to the elementary discontinuity of energy. The energy that serves as information carrier is in fact modulated in two ways: artificially, by the signal that is to be transmitted; essentially, by virtue of its physical nature, by the elementary discontinuity. This elementary discontinuity appears when the average energy level is of an order of magnitude that is barely superior to instantaneous variations due to the elementary discontinuity of energy; artificial modulation then conflates itself with this essential modulation, with this white noise or fog that superimposes itself on the transmission; it is not a question here of a harmonic distortion, for this is a modulation that is independent from that of the signal, and not simply a deformation or impoverishment of the signal. Now, in order to diminish the background noise, one can diminish the bandwidth, which also diminishes the informational efficiency of the channel under consideration. A compromise must be struck that preserves an informational efficiency sufficient for practical needs and an energetic efficiency that is sufficiently high to keep background noise at a level where it does not trouble the reception of the signal.

And yet this antagonism, which is barely acknowledged in the recent work going on in the philosophy of information technics, marks the non-univocal aspect of the notion of information. Information is, in a sense, that which can be infinitely varied, that which, in order to be transmitted with the least possible loss, requires a sacrifice of energy efficiency so as to avoid any reduction of the range of possibilities. The most faithful amplifier is the one with a very uniform efficiency independent of the scale of frequencies; it favors none of them, imposes no resonance, no stereotypy, no pre-established regularity onto the open series of varied signals that it must transmit. But information, in another sense, is that which, in order to be transmitted, must be above the level of pure random phenomena, such as white noise or thermal agitation; information is then that which possesses a regularity, a localization, a defined domain, a determined stereotypy through which information distinguishes itself from pure chance. When the level of background noise is high, the information signal can still be saved if it has a certain law, in other words, if it offers a certain predictability of the unfolding of the temporal series of the successive states that constitute it. In television, for instance, the fact that the frequency of the timebase is well determined in advance allows for the synchronization timing pips to be extracted from an equally important background noise by blocking the synchronization devices nine-tenths of the time, and by unblocking them for just a brief instant (for instance, a millionth of a second) when the synchronization timing pip must set in, by virtue of the previously defined law of recurrence (this is the phase comparison device, used for long distance receptions). The reception of synchronization signals, in turn, must indeed be treated as information. But this information is more easily extracted from background noise because the perturbing action of the background noise can be limited to a very small fraction of the total time, thereby rejecting as insignificant all the manifestations of background noise that fall outside this instant. This device is of course ineffective against a parasitic signal, which itself obeys a recurrent law with a period very close to the period intended for signal reception. There are thus two aspects of information, which distinguish themselves technically through the opposed conditions necessary for their transmission. In one sense, information is that which brings about a series of unpredictable, new states, not belonging to any series that could be defined in advance; it is thus that which requires an absolute availability of the information channel regarding all the aspects of modulation that it bears; the information channel itself must not contribute any predetermined form of its own, nor be selective. A perfectly faithful amplifier would have to be able to transmit all frequencies and amplitudes. In this sense, information has certain aspects in common with purely contingent, lawless phenomena, such as the movements of thermal molecular agitation, radioactive emissions, discontinuous electronic emissions in the thermoelectric or photoelectric effect. This is why a very faithful amplifier[20] has greater background noise than an amplifier with a smaller bandwidth, because it uniformly amplifies the white noises produced in its diverse circuits by diverse causes (by thermal effect in the resistances, by discontinuity of electronic emission in the tubes). Noise, however, has no signification whereas information has signification. By contrast, information distinguishes itself from noise because it can be assigned a certain code, a relative uniformization; in any case where noise cannot be directly reduced below a certain level, a reduction of the margin of indeterminacy and unpredictability of the information signals is performed; such is the case, mentioned above, of the reception of synchronization signals by a phase comparator. What is reduced here, is the margin of temporal indeterminacy: it is assumed that the signal will produce itself at a certain moment in a temporal interval equal to a minimal fraction, perfectly determined by its phase, of the period of the recurrent phenomenon. The device can be tuned all the more finely as the stability of the transmitter and the stability of the receptor increase. The greater the predictability of the signal, the easier it is for it to be distinguished from the chance phenomenon called background noise. The same holds true for the reduction of the frequency band: when a circuit can no longer transmit speech, because of a high level of background noise, one can use a transmission of signals in a single frequency, as is done with Morse code; at reception, a filter adapted to the unique frequency of transmission only allows for those sounds to pass whose frequency falls within this narrow band; a low level of background noise then comes through, all the more reduced as the received band narrows, i.e., as the resonance becomes sharper.

189

This opposition represents a technical antinomy that poses a problem for philosophical thought: information is *like* the chance event, but it nevertheless distinguishes itself from it. An absolute stereotypy, excluding all novelty, also excludes all information. And yet, in order to distinguish information from noise, one takes an aspect of the reduction of the limits of indeterminacy as a basis. If the time bases were truly incorruptible like Leibniz's monads, then one could reduce the synchronization time of the oscillator as much as desired; the informing role of the synchronizing pulse would entirely disappear, because there would be nothing to synchronize: the synchronization signal would have no aspect of unpredictability with respect to the oscillator to be synchronized; in order for the informational

---

20. With a large bandwidth.

nature of the signal to subsist, a certain margin of indeterminacy must subsist. Predictability is the ground receiving this supplementary precision, distinguishing it in advance from pure chance in a great number of cases, partially preforming it. Information is thus halfway between pure chance and absolute regularity. One can say that form, conceived as absolute spatial as well as temporal regularity, is not information but a condition of information; it is what receives information, the

190 *a priori* that receives information. Form has a function of selectivity. But information is not form, nor is it a collection [*ensemble*] of forms; it is the variability of forms, the influx of variation with respect to a form. It is the unpredictability of a variation of form, not pure unpredictability of all variation. We would thus be led to distinguish three terms: pure chance, form, and information.

However, to this day, the new phase of the philosophy of technics, which followed after the phase that was contemporary with thermodynamics and energetics, has not made a clear distinction between *form* and *information*. There is, in effect, an important gap between the living thing and the machine, and consequently between man and machine, which comes from the fact that the living thing needs information, while the machine essentially uses forms, and is so to speak constituted with forms. Philosophical thought will not be able to grasp the sense of coupling between man and machine unless it manages to elucidate the true relation that exists between form and information. The living transforms information into forms, the *a posteriori* into *a priori*; but this *a priori* is always oriented toward the reception of information to be interpreted. The machine on the contrary has been built according to a certain number of schemas, and it functions in a determinate way; its technicity, its functional concretization at the level of the element are determinations of forms.

The human individual thus appears as having to convert the forms deposited into machines into information; the operating of machines does not give rise to information, but is simply an assemblage and a modification of forms; the functioning of a machine has no sense, and cannot give rise to true information signals for another machine; a living being is required as mediator in order to interpret a given functioning in terms of information, and in order to convert it into the forms for another machine. Man understands machines; for there to be a true technical

191 ensemble man has to play a functional role between machines rather than above them. It is man who discovers significations: signification is the meaning [*sens*] that an event takes on with respect to already existing forms; signification is what makes an event have value as information.

This function is complementary with the function of the invention of technical individuals. Man, interpreter of machines, is also the one who has, on the basis of his schemas, founded the rigid forms that enable the machine to function. The machine is a deposited fixed human gesture that has become a stereotypy and the power to restart. A rocker switch with two fixed states* has been thought out and built once; man represented its functioning to himself a limited number of times, and now the rocker performs its equilibrium-reversing operation indefinitely. It perpetuates the human operation that constituted it in a determinate activity; a certain transition has been carried out, through construction, from a mental to a physical functioning. There is a veritable and profound dynamic analogy between the process through which man thought up the rocker switch and the physical process of functioning of this rocker switch once built. A relation of iso-dynamism exists between man who invents and the machine that functions, which is more essential than the Gestalt psychologists had imagined in order to explain perception, calling it an isomorphism. The analogical relation between machine and man is not at the level of corporeal functioning; the machine neither nourishes itself nor perceives, nor rests, and cybernetic literature falsely exploits the appearance of analogy. The true analogical relation is in fact between the mental functioning of man and the physical functioning of the machine. These two ways of functioning are not parallel within everyday life, but rather within invention. To invent is to make one's thought function as a machine might function, neither according to causality, which is too fragmentary, nor according to finality, which is too uni- 192 tary, but according to the dynamism of lived functioning, grasped because it is produced, accompanied in its genesis. The machine is a being that functions. Its mechanisms concretize a coherent dynamism that once existed in thought, which were that thought. During invention, the dynamism of thought converted itself into functioning forms. Inversely, the machine, in functioning, is subject to or produces a certain number of variations around the fundamental rhythms of its functioning, arising from its definite forms. These variations are what are significant, and they are significant with respect to the archetype of functioning, which is that of thought in the process of invention. One has to have invented or reinvented the machine if the machine's variations of functioning are to become information. The noise of an engine in itself does not have value as information; it takes on this value through its variation in rhythm, the change of its frequency or tone, the alteration of its transients which translate a modification of its functioning with respect to the functioning that results from invention. When the correlation that exists between machines is purely causal, then it isn't necessary for a human being to intervene as the mutual interpreter of the machines. But this role is necessary when the machines comprise regulation; a machine that has regulation is in effect a machine that harbors a certain margin of indeterminacy in its functioning; it can, for instance, go fast or slow. Henceforth, its variations in speed become significant and can take into account what is happening outside the machine, in the technical ensemble. The more automatized the machines are, the more restricted its possible variations in speed are; they can thus go unperceived; but what is happening here is in fact what is happening for a very stable oscillator that is synchronized by another even more stable oscillator: the oscillator can continue to receive infor-

193   mation as long as it is not rigorously stable, and despite the fact that the margin of indeterminacy of its functioning is reduced, synchronization still has meaning [*sens*] inside this margin of indeterminacy. The synchronization pulse has meaning when it intervenes as a very slight variation on this temporal form of recurrence of the states of functioning. In the same way, the reduction of the indeterminacy of its functioning does not isolate the machines from one another; it renders the variation in signification, which has value as information, more rigorous and refined. But it is always with respect to the essential schemas of the invention of the machine that these variations have a sense.

The notion of a perfect automaton is a notion that is obtained by confronting a limit, and so it harbors something contradictory: *the automaton is supposed to be a machine so perfect that the margin of indeterminacy of its functioning would be null, but which would be able nevertheless to receive, interpret, or emit information.* And yet, if the margin of indeterminacy of functioning is null, then there is no longer any possible variation; the functioning repeats indefinitely, and consequently this iteration has no signification. Information is maintained throughout automatization only because the fineness of the signals increases with the reduction of the margin of indeterminacy, which means that the signals conserve their signifying value, even if this margin of indeterminacy becomes extremely narrow. If, for instance, oscillators are stable to about a thousandth in frequency variation, then the synchronization pulses, whose possible phase-rotation would vary over time at about ten percent, or which would have no steep incline and a variable duration, would have only a low informational value for synchronization. To synchronize oscillators that are already very stable, tiny perfectly cut pulses are used whose phase-angle is

194   rigorously constant. Information is all the more significant, or rather a signal has all the more informational value as it acts in concordance with an autonomous form of the individual who receives it; hence, when the natural frequency of an oscillator that is to be synchronized is far from the frequency of the synchronization

pulses, synchronization does not occur; on the contrary, synchronization occurs for signals that become weaker as the autonomous frequency and the pulse frequency get closer to each other. Nevertheless, this relation must be more carefully interpreted: for recurrent pulses to synchronize an oscillator, these pulses must arrive at a critical period of functioning, namely the one which immediately precedes the reversal of equilibrium, which is to say just before the beginning of a phase; the synchronizing pulse arrives as a very small additional quantity of energy that accelerates the passage to the next phase, at a moment in which this passage was not yet fully accomplished; the pulse *triggers*. It is for this reason that the greatest fineness of the synchronization, the highest sensitivity, is obtained when the autonomous frequency would be ever so slightly lower than the synchronizing frequency. With respect to this form of recurrence, the pulses with a very slight advance take on a meaning [*sens*], and carry information. The moment in which the oscillator's equilibrium will be reversed is that in which a metastable state is created, with an accumulation of energy.

It is this existence of critical phases that explains the difficulty of synchronizing a way of functioning that does not offer an abrupt reversal of states: a sinusoidal oscillator synchronizes less easily than a relaxation oscillator*; the margin of indeterminacy is effectively less critical in the functioning of a sinusoidal oscillator; its functioning can be modified at any moment during the course of its period; in a relaxation oscillator, on the contrary, indeterminacy is accumulated at each end of the cycle, rather than being spread over the whole duration of the cycle; when the 195 equilibrium is reversed, the relaxation oscillator is no longer sensitive to the pulse it receives; but when it is on the tipping point, it is extremely sensitive; the sinusoidal oscillator, on the contrary, is sensitive throughout the phase, but in a mediocre way.

The existence of a margin of indeterminacy in machines must therefore be understood as the existence of a certain number of critical phases of functioning; the machine that can receive information is the one that temporarily localizes its indeterminacy in instants that are sensitive and rich with possibilities. This structure is one of decision, but it is also that of the relay. The machines that can receive information are the ones that localize their indeterminacy.

This notion of the localization of decisions within functioning is not absent from books on cybernetics. But what this study lacks is the notion of the reversibility of the reception of information and the emission of information. If a machine presents a functioning with critical phases, such as those of the relaxation oscillator, then it can emit information as well as receive it; a relaxation oscillator thus emits pulses, as a result of its discontinuous functioning, which can serve to synchronize another relaxation oscillator. If one couples two relaxation oscillators to each other, these two oscillators synchronize in such a way that one cannot say which one synchronizes and which one is synchronized; in fact, they synchronize each other, and the ensemble functions as a single oscillator, with a single period that is slightly different from the periods proper to each one of the oscillators.

It can appear too simple to oppose open machines and closed machines, in the sense Bergson gives these two adjectives.[21] And yet this difference is real; the existence of a regulation in a machine leaves the machine open insofar as it localizes the critical periods and the critical points, in other words those on the basis of which the energetic channels of the machine can be modified, changing characteristics. The individualization of the machine goes hand in hand with this separation of forms and critical elements; a machine can be in relation with the exterior insofar as it possesses critical elements; the existence of these critical points in the machine, in turn, justifies the presence of man: the rate [*régime*] of the machine can be modified by information coming from the outside. A calculating machine is therefore not only, as one generally says, a set [*ensemble*] of rocker switches. It is true that the calculating machine has a great number of determinate forms, the forms of the functioning of the series of rocker switches, representing a series of operations of additions. But if the machine consisted in this alone, then it would be useless, because it wouldn't be capable of receiving any information. In fact, it also has what one could call the system of schemas of decisions; before operating the machine, it has to be *programmed*. Even with the multi-vibrator that provides the pulses and the series of rockers that do the adding, there still wouldn't be a calculating machine. It is the existence of a certain degree of indeterminacy that makes calculation possible: the machine contains a set [*ensemble*] of selectors and of commutations that are commanded by programming. Even in the simplest case, that of a scale composed of rockers and counting pulses, such as those used after the Geiger-Müller tube-counters, there is a degree of indeterminacy in its functioning; the Geiger tube under voltage is in the same state as a relaxation oscillator at the instant in which it will start a new phase, or as a multi-vibrator in the instant in which it will switch by itself. The only difference is that this metastable state (corresponding to a constant voltage in the Geiger-Müller tube) prolongs itself in the tube in a durable way until an additional energy triggers ionization, whereas, in the relation oscillator or multi-vibrator, this state is transitory, as a result of the

---

21. Henri Bergson, *The Two Sources of Morality and Religion*, trans. R. Ashley Audra (Notre Dame, IN: University of Notre Dame Press, 1977). [TN]

continuation of activity in the resistance circuits and the capacitances external to the electronic tube or thyratron.

This margin of indeterminacy can be found in all the varying types of devices that can transmit information. A continuous relay, such as a thermo-electronic or crystalline triode, can transmit information because the existence of potential energy defined at the limits of the supply circuit is not enough for determining the quantity of effective and actual energy that is sent into the outlet circuit: this open relation of possibility in the actualization of an energy is closed only by way of the additional condition of the arrival of information onto the controller. A continuous relay can be defined as a transducer, in other words as resistance that can be modulated by an information that is external to the potential energy and to the actual energy: this resistance can be modulated by information that is external to both the potential energy and the real energy. And still the notion of "resistance that can be modulated" is too vague and inadequate; if this resistance was effectively a true resistance, it would belong to the domain of actualization of the potential energy. Yet, in a perfect transducer, no energy is actualized; nor is any energy put in reserve: the transducer belongs neither to the domain of potential energy, nor to the domain of actual energy; it is truly the mediator between these two domains, but it is neither a domain of the accumulation of energy, nor a domain of actualization: it is a margin of indeterminacy between these two domains, that which brings potential energy to its actualization. It is during the course of this passage from potential to actual that information comes into play; information is the condition of actualization.

This notion of transduction, in turn, can be generalized. Presented in its purest state in transducers of different kinds, it exists as a regulative function in 198 all machines having a certain margin of localized indeterminacy in their functioning. The human being, and the living being more generally, are essentially transducers. The elementary living being, the animal, is itself a transducer when it stores chemical energies and then actualizes them during the course of different vital operations. Bergson illuminated this function of the living, which constitutes energetic potentials and expends them briskly; but Bergson was here concerned with showing a function of temporary contraction that would be constitutive of life; the relation, however, between the slowness of accumulation and the instantaneous briskness of actualization does not always exist; the living thing can actualize its potential energy slowly, as in thermal regulation or muscle tonicity; what is essential is not the difference of temporal operating speeds of potentialization and actualization, but the fact that the living thing intervenes between potential energy and this actual energy as a transducer; the living thing is *that which modulates*, that in which there is modulation, and not a reservoir of energy or an effector. Nor is it enough to say: the living assimilates; assimilation is a source of potential energy that can be liberated and actualized in the functions of transduction.

Man's relation with machines takes place at the level of the functions of transduction. It is in fact very easy to construct machines that ensure a much greater accumulation of energy compared to that which man can accumulate in his body; it is equally possible to use artificial systems that constitute effectors that are superior to those of the human body. But it is very difficult to construct transducers comparable to the living thing. Indeed the living thing is not exactly a transducer like those that can be found in machines; it is that and something more; mechanical transducers are systems with a margin of indeterminacy; information is that which adds determinacy. But this information must be given to the transducer; it cannot invent it; it is given to it by a mechanism that is analogous to that of perception in the living, for instance by a signal coming from the manner in which the effector functions (the gauge on the output shaft of a heat engine). On the contrary, the living thing has the capacity to give itself information, even in the absence of all perception, because it possesses the capacity to modify the forms of the problems to be resolved; for the machine, there are no problems, only data that modulate the transducers; several transducers that act upon one another according to commutable schemas, such as Ashby's homeostat, do not constitute a problem solving machine: the transducers in a reciprocal relation of causality are all *within the same time*; they condition one another in actuality [*dans l'actuel*]; they are never confronted with a problem, something thrown down before them, something that is in front of them and that they will have to step or leap over. To solve a problem is to be able to step over it, to be capable of recasting the forms that are given within the problem and in which it consists. The solution of real problems is a vital function presupposing a recurrent mode of action that cannot exist in the machine: the recurrence of the future with respect to the present, of the virtual with respect to the actual. There is no true virtuality in a machine; the machine cannot reform its forms in order to solve a problem. When Ashby's homeostat switches by itself during the course of its functioning (for one can attribute to this machine the faculty of acting on its own selectors), a jump in characteristics that eliminates all prior functioning occurs; at each instant the machine exists in actuality, and the faculty to apparently change its forms is not very efficient, because nothing is left of the previous forms; it all happens as if there were a new machine; each operation is momentary; when the machine changes form by switching, it does not switch in order to have this other form oriented toward solving the problem; there is no modification of forms that would be oriented by the presentiment of a problem to be resolved; the virtual does not act upon the actual, because the virtual, insofar as it is virtual, cannot play a role for the machine. It can only react to something that is positively given, actually done. The living thing has the faculty to modify itself according to the virtual: this faculty is the sense of time, which the machine does not have because it does not live.

Technical ensembles are characterized by the fact that in them a relation between technical objects takes shape at the level of the margin of indeterminacy of each technical object's way of functioning. This relation between technical objects is of a problematic type, insofar as it puts indeterminacies into correlation, and for this reason it cannot be taken on by the objects themselves; it cannot be calculated, nor be the result of a calculation; it must be thought, posed as a problem by a living being and for a living being. One could express what we have called a coupling between man and machine by saying that man is responsible for machines. This responsibility is not that of the producer insofar as the produced thing comes from him, but that of a third party, a witness to a difficulty that only he can resolve because he alone has the power to think; man is the witness to the machines and represents them in relation to one another; machines can neither think nor experience [*vivre*] their mutual relation; they can only act upon one another in actuality, according to causal schemas. Man as witness to machines is responsible for their relation; the individual machine represents man, but man represents the ensemble of machines, for there is not one machine of all machines, whereas there can be a thought that encompasses all machines.

One can call a technological attitude that which compels man to look after not only the utilization of a technical being, but after the correlation of technical beings among each other. The current opposition between culture and technics comes from the fact that the technical object is considered identical with the machine. Culture does not understand the machine; it is inadequate to technical reality because it considers the machine to be a closed block, and considers mechanical functioning to be an iterative stereotypy. The opposition between technics and culture will last until culture discovers that each machine is not an absolute unit, but only an individualized technical reality that is open according to two paths: that of the relation to elements, and that of the inter-individual relations within the technical ensemble. The role culture has assigned to man alongside the machine is at odds with technical reality; it assumes that the machine is substantialized, materialized, and consequently devalued; the machine is in fact less consistent and

201

less substantial than culture assumes; it does not relate to man as a single block, but through the free plurality of its elements, or the open series of its possible relations with other machines within the technical ensemble. Culture is unjust toward the machine, not only in its judgments or its preconceptions, but at the very level of knowledge: the cognitive intention of culture toward the machine is substantializing; the machine is closed up in this reductive vision that considers it to be perfect and finished in itself, that makes it coincide with its actual state, with its material determinations. With respect to the art object, such an attitude would consist in reducing a painting to a certain expanse of dried and cracked paint on a stretched canvas. In regard to the human being, the same attitude would consist in reducing the subject to a fixed set of vices and virtues, or character traits.[22]

202   To reduce art to art objects, to reduce humanity to a series of individuals that are mere carriers of character traits, is to act as one does when reducing technical reality to a collection of machines: yet, in the first two cases, this attitude is judged crude, in the second case, it passes for being in conformity with the values of culture, although it operates with the same destructive reduction as it does in the first two cases. Except that it operates by making an implicit judgment through knowledge itself. It is the very notion of the machine that is already distorted, like the representation of the foreigner [*étranger*] in group stereotypies.

However, it is not the foreigner as foreigner who can become the object of cultured thought, but only the human being. The stereotype of the foreigner cannot be transformed into a just and adequate representation unless the relation between the being who judges and the one who is the foreigner is diversified and is multiplied in order to acquire a multiform mobility that confers upon it a certain consistency, a definite power of reality. A stereotype is a two-dimensional representation, like an image, without depth and without plasticity. For the stereotype to become a representation, the experiences of the relationship with the foreigner must be multiple and varied. The foreigner is no longer foreign, but other, when there are foreign beings not only with respect to the subject who judges, but also with respect to other foreigners; the stereotype falls away when this relation of man to the foreign is known in its entirety between other people, rather than enclosing the subject and the foreigner within an asymmetrical, immutable mutual situation. In the same manner, the stereotypes relating to the machine cannot change unless the relation between man and machine (an asymmetrical relation for as long as it is lived as an exclusive relation) can be seen objectively in the process of producing itself between terms that are independent of the subject, between technical objects.

22. This reductive attitude can also exist toward an entire region (regionalism).

In order for the representation of technical contents to be incorporated into cul- 203
ture, an objectivation of the technical relation must exist for man.

The predominant and exclusive attention to a machine cannot lead to the discovery of technicity, any more than the relation with a single sort of stranger or
foreigner can allow one to penetrate the interiority of their way of life, and to know
it according to culture. Even interaction with several machines is not enough, any
more than successive interactions with several foreigners; these experiences only
lead to xenophobia or to xenophilia, which are opposite but equally passionate
attitudes. In order to consider a foreigner through culture, one needs to have objectively seen the relation whereby two beings are foreigners to one another play out.
Likewise, if a single technics does not suffice to offer cultural content, neither does
a polytechnics; it merely engenders the tendency toward technocracy or the refusal
of technics as a whole.

*IV. – Philosophical thought must carry out the integration of technical reality into
universal culture, by founding a technology*

The advent of the conditions allowing man to see the technical relation functioning
in an objective way is the prime condition for the incorporation of the knowledge
of technical reality and of the values implied by its existence into culture. Now,
these conditions are realized in the technical ensembles employing machines that
have a sufficient degree of indeterminacy. For man, the action of having to intervene as a mediator in this relation between machines grants him a situation of
independence in which he can acquire a cultural vision of technical realities. The
engagement in the asymmetrical relation with a single machine cannot provide the 204
necessary distance for what one might call technical wisdom. Only a situation in
which there is a concrete link with machines and a responsibility toward them, but
which is liberated vis-à-vis each one taken individually, can provide this serenity
of having technical awareness. Just as literary culture, in order to constitute itself,
needed the wise individuals who had lived and contemplated the inter-human
relation with a certain distance that gave them a serenity and a depth of judgment
while nevertheless maintaining an intense presence among human beings, technical culture cannot constitute itself without developing a certain sort of wisdom,
which we will call technical wisdom, in men who feel their responsibility toward
technical realities, but who remain disengaged from the immediate and exclusive

relationship with a particular technical object. It is rather difficult for a worker[23] to know technicity through the aspects and modalities of his daily work on a machine. It is also difficult for a man who is the owner of machines and who considers them productive capital to know their essential technicity. It is the mediator of the relation between machines alone who can discover this particular form of wisdom. Such a function, however, does not yet have a social place; it would be that of the production planning engineer if he wasn't preoccupied by immediate output, and governed by a finality external to the operating system [*régime*] of machines which is that of productivity. The function whose basic lines we are attempting to draw would be that of a psychologist of machines, or of a sociologist of machines — what we might call a mechanologist.

A sketch of this role can be found in Norbert Wiener's founding intention for cybernetics, this science of command and communication in the living being and the machine. The meaning of cybernetics has been poorly understood, for this eminently new endeavor has been reduced and judged according to old notions or tendencies. In France, cybernetic research, which makes the assumption that there is a unity between information theory and the study of command schemas and self-regulation, has split into two divergent branches, that of information theory with Louis de Broglie and the team publishing its work in the journal *Revue d'Optique*, and that of researchers on automatism, with engineers like Albert Ducrocq, representing the technicist and technocratic tendencies. But it is the connection between these two tendencies that would enable the discovery of the values implied in technical realities and their incorporation into culture. Information theory is indeed of a scientific order: it employs operational modes close to those employed by thermodynamic theory. Ducrocq's technicism, on the contrary, seeks within the functioning of automatic machines the example of a certain number of functions that allow one to interpret other types of realities by analogy with automatism. In particular, the theory of the mechanisms of self-regulation allows us to sketch a hypothesis for explaining the origins of life. Or else it is the principal mental operations, or certain nervous functions that are explained by analogy in this way. Similar analogies, even when they are not arbitrary, in fact merely indicate that there are ways of functioning that are in common to both living things and machines. They leave the problem of nature itself and of its ways of functioning untouched: this technicism is a phenomenology rather than a deepening of the search for the nature of the schemas and conditions that govern their implementation.

---

23. The neutral term *operator* would be appropriate.

It is of course possible not to accept the way Norbert Wiener characterizes information, and the essential postulate of his book, which consists in affirming that   206
information is opposed to background noise in the same way that negative entropy
is opposed to entropy as defined by thermodynamics. However, even if this opposition of a divergent determinism to a convergent determinism does not account
for all of technical reality and its relation with life, this opposition contains within
itself an entire method for the discovery and for the definition of a set of values
that are implied in technical ways of functioning and in the concepts by means of
which one can think them. But it is possible to prolong Norbert Wiener's reflection. At the end of his book, the author wonders how the concepts he has defined
might be used for the organization of society. Norbert Wiener observes that large
groups contain less information than small groups, and he explains this fact by
the tendency of the human elements that are the least "homeostatic" to occupy
leading functions in large groups; according to Norbert Wiener, the quantity of
information contained in a group would be, on the contrary, proportionate to
the degree of perfection of the group's homeostasis. The fundamental moral and
political problem would then consist in asking oneself how one can put individuals
representing homeostatic functions at the heads of groups. But Norbert Wiener
claims that none of the individuals who understand the value of homeostasis and
who also understand information are capable of taking power; and all the cyberneticians together are confronted with men who preside over collective destinies,
like mice that want to hang a bell around the cat's neck (*Cybernetics*[24]). The forays
made by the author among trade union leaders filled him with a bitterness that
calls to mind Plato describing his disappointments in his *Seventh Letter*. And yet,
one can try to find a very different mediation between the understanding of technics and the force that governs human groups from the one Wiener contemplates.
For it is difficult to make philosophers kings and kings philosophers. It often hap-   207
pens that philosophers who have become kings cease to be philosophers. The true
mediation between technics and power cannot be individual. It can be realized
only through the mediation of culture. For there is something that allows man to
govern: the culture he has received; it is this culture that gives him significations
and values; it is culture that governs man, even if this man in turn governs other
men and machines. Now, this culture is elaborated by the great mass of those who
are governed; to the extent that the power exerted by a man does not strictly speaking come from himself, but crystallizes and concretizes in him; it comes from the

---

24. Norbert Wiener, *Cybernetics: or Control and Communication in the Animal and the Machine* (Cambridge, MA: MIT
Press, 1948), 162.

governed men and it returns there. There is a sort of recurrence there.

In a time when the development of technics was poor, the elaboration of culture by governed men was enough for the government to think the problems of the group as a whole: because it went from human group to human group via the government, the recurrence of causality and information was complete and accomplished. But this is no longer true: the basis of culture is still exclusively human; it is elaborated by the group of men; however, having gone through government, it returns and applies itself to the human group on the one hand and to machines on the other: machines are ruled by a culture that has not been elaborated according to them, and from which they are absent; this culture is inadequate for them and does not represent them. If the totality of reality escapes the man who governs, then this is because the basis of this reality is still exclusively human. It is culture that is regulative and in which the circular link of causality between governing and governed consists: its starting point and its fulfillment are the governed. The lack of social homeostasis comes from the fact that there is an aspect of governed reality that is not represented in this regulating relation that is culture.

Therefore, the task of the technologist is to be the representative of the technical beings whereby culture is elaborated: writers, artists, and, very generally, those called cynosures in psycho-sociology. This is not about mechanizing society, through the integration of an adequate representation of technical realities to culture. There is no reason to consider society as an unconditional domain of homeostasis. Norbert Wiener appears to admit a postulate of values that is unnecessary, which is to say that a good homeostatic regulation is the ultimate purpose of societies, and the ideal that must animate every act of government. In fact, just as the living [*le vivant*] is grounded in homeostases so as to develop itself and to continue its coming-into-being, rather than remaining perpetually in the same state, so too in the act of government there is a force of absolute advent, which is based on homeostases, but which surpasses them and makes use of them. The integration of a representation of technical realities into culture, through an elevation and enlargement of the technical domain, must put the problems of finality in their place, as technical problems, which are wrongly considered ethical and sometimes religious problems. The incompleteness of technics sacralizes problems of finality and enslaves men with respect to ends that he represents to himself as absolutes.

For this reason, it is not only technical objects that need to be known at the level of what they currently are, but the technicity of these objects as a mode of relation of man with the world among other modes such as the religious and the aesthetic. Taken on its own, technicity tends to become domineering and to answer each and every problem, as it does today through the system of cybernetics. In fact, in order for it to be accurately known, according to its essence, and rightly integrated into culture, technicity must be known in its relation with the other modes of man's being in the world. No inductive study, starting from the plurality of technical objects, can discover the essence of technicity: it is therefore a direct examination of technicity according to a genetic method that must be attempted, by employing a philosophical method.

PART III
THE ESSENCE OF TECHNICITY

The existence of technical objects and the conditions of their genesis pose a question for philosophical thought which it cannot resolve through the simple consideration of technical objects themselves: what is the sense of the genesis of technical objects with respect to the whole of thought, of man's existence, and of his manner of being in the world? The fact that there is an organic aspect of thought and of the mode of being in the world obliges one to assume that the genesis of technical objects has repercussions on other human productions, on the attitude man has toward the world. But this is only a lateral and very imperfect way of posing the problem that the manifestation of technical objects leads us to confront, as a reality that is subject to genesis and whose genuine essence resides only in the lines of this genesis. Indeed, nothing proves what we have here is an independent reality, which is to say the technical object taken as a definite mode of existence.

If the mode of existence is definite because it derives from a genesis, then this genesis that engenders objects is perhaps not only the genesis of objects, or even the genesis of technical reality: perhaps it originates even further back and constitutes a limited aspect of a larger process, and continues perhaps to engender other realities after having led to the appearance of technical objects. It is thus the genesis of all of technicity that needs to be understood, that of objects and that of non-objectified realities, and the entire genesis implicating man and the world, of which the genesis of technicity is perhaps only a small part, shouldered and balanced by other geneses, that are prior, posterior or contemporary, and correlative with that of technical objects.

We therefore need to move toward a generalized genetic interpretation of the

relation between man and the world in order to grasp the philosophical importance of the existence of technical objects.

The very notion of genesis, however, deserves to be made more precise: the word
genesis is taken here in the sense defined in the study on *Individuation in the Light
of the Notions of Form and of Information*, as the process of individuation in its generality. There is genesis when the coming-into-being of a system of a primitively
oversaturated reality, rich in potential, greater than unity and harboring an internal
incompatibility, constitutes for this system the discovery of compatibility, a resolution
through the advent of structure. This structuration is the advent of an organization
that is the basis of an equilibrium of metastability. Such genesis opposes itself to the
degradation of the potential energies contained in a system through the passage to a
stable state from which transformation is no longer possible.

The general hypothesis we are making about the sense of the coming-into-being of man's relation with the world consists in considering the whole [*l'ensemble*]
formed by man and the world as a system. This hypothesis, however, is not limited to the affirmation that man and the world form a vital system, comprising
the living thing and its milieu; evolution could indeed be considered an adaptation, i.e., a quest for a stable equilibrium of the system through a reduction of the
gap between the living thing and the milieu. The notion of adaptation, however,
together with the notion of function and of functional finality to which it is linked,
would lead us to see the coming-into-being of the relation between man and the
215   world as tending toward a state of stable equilibrium, which does not appear to
be correct in man's case, perhaps no more than it is for any living being. If one
wanted to preserve a vitalist foundation for this hypothesis of genetic coming-into-
being, then one could call upon the notion of *élan vital* presented by Bergson. And
indeed, this notion is excellent for demonstrating what is lacking in the notion of
adaptation in order for it to enable an interpretation of the coming-into-being of
life [*devenir vital*], but it is not compatible with it, and an antagonism without possible mediation subsists between adaptation and the *élan vital*. It seems that these
two opposed notions, as the couple they form, can be replaced, by the notion of
the individuation of oversaturated systems, conceived as successive resolutions of
tensions through the discovery of structures at the heart of a system rich in potential. Tensions and tendencies can be conceived as really existing in a system: the
potential is one of the forms of the real, as completely as the actual. The potentials
of a system constitute its power of coming-into-being without degradation; they
are not the simple virtuality of future states, but a reality that pushes them into
being. Coming-into-being is not the actualization of a virtuality or the result of a

conflict between actual realities, but the operation of a system with potentials in its reality: coming-into-being is a series of spurts of structurations of a system, or of successive individuations of a system.

The relation of man to the world is not a simple adaptation, governed by a law of self-regulating finality that would find ever increasingly stable states of equilibrium; on the contrary, the evolution of this relation, in which technicity participates among other modes of being, manifests an ability to evolve that grows at each stage, discovering new forms and forces capable of making it evolve even more, rather than stabilizing it and making it tend toward more and more limited fluctuations; the very notion of finality, applied to this coming-into-being, appears inadequate, since one can indeed find limited finalities within this coming-into-being (search for food, defense against destructive forces), but there is no unique 216 and superior end that one could superimpose on all aspects of evolution in order to coordinate them and account for their orientation through the search for an end that would be superior to all particular ends.

This is why it is not forbidden to call upon a hypothesis that intervenes with a more primitive genetic schema than that of the opposed aspects of adaptation and *élan vital*, and enclosing both of them as abstract limit cases: namely that of the successive stages of an individuating structuration, going from metastable state to metastable state by means of successive inventions of structures.

The technicity that manifests itself in the use of objects can perhaps be conceived as appearing in a structuration that provisionally resolves the problems posed by the primitive and original phase of man's relationship to the world. One can call this first phase the *magical phase*, taking this word in its most general sense, and considering the magical mode of existence as the pre-technical and pre-religious mode, and so as immediately above the relationship that is simply between the living thing and its milieu. The magical mode of the relation with the world is not devoid of all organization: on the contrary, it is rich in implicit organization, attached to the world and to man; in the magical mode the mediation between man and the world is not yet concretized and constituted as standing apart, by means of specialized objects or human beings, but this mediation does exist functionally in the most elementary of all structurations, which is also the first: that from which erupts the distinction between figure and ground in the universe. Technicity appears as a structure that resolves an incompatibility: it specializes the figural functions, while religions on their side specialize the functions of ground; the original magical universe, which is rich in potentials, structures itself by splitting in two. Technicity appears as one of the two aspects of a solution given to the problem of man's relation with the world, the other simultaneous and correlated aspect being the institution of definite religions. However, coming-into-being does not stop at the discovery of technicity: from being a solution, technicity once more becomes a problem when it reconstitutes a system via an evolution that leads from technical objects to technical ensembles: the technical universe is saturated then over-saturated in turn, at the same time as the religious universe, just as had happened with the magical universe. The inherence of technicity to technical objects is provisional; it constitutes only a moment of genetic coming-into-being.

Now, according to this hypothesis, technicity must never be considered an isolated reality, but as part of a system. It is a partial reality and a transitory reality, both the result and principle of genesis. Being the result of an evolution, technicity is also the depositary of a capacity to evolve, precisely because, as solution to a first problem, it has the capacity to be a mediation between man and the world.

This hypothesis would entail two consequences: first, the technicity of objects or of thought cannot be considered as a complete reality or as a mode of thinking having its own independent truth; any form of thought or any mode of existence engendered by technicity would need to be complemented and balanced out by another mode of thought or existence coming from the religious world.

Secondly, as the emergence of technicity marks a break within and a splitting in two of the primitive magical unity, technicity, like religiosity, is heir to a capacity for evolutionary divergence; in the coming-into-being of man's mode of being in the world, this force of divergence must be compensated by a force of convergence, by a relational function maintaining unity despite this divergence; the splitting of the magical structure wouldn't be viable unless a function of convergence stood in opposition to the powers of divergence.

These two reasons make it necessary to study where technicity comes from, what it results in, and what relations it maintains with man's other modes of being, which is to say how it opens itself to functions of convergence.

Now, the general sense of coming-into-being would be the following: the different forms of thinking and of being in the world diverge as soon as they appear, i.e., when they are not saturated; then they re-converge when they are saturated and tend to structure themselves through new splits. The functions of convergence can take place by virtue of the over-saturation of the evolutionary forms of being in the world, at the spontaneous level of aesthetic thought and at the reflected level of philosophical thought.

Technicity becomes oversaturated by once again incorporating the reality of the world to which it is applied; religiosity, on the other hand, becomes oversaturated by incorporating the reality of human groups whose primitive relation with the world it mediates. Thus oversaturated, technicity splits into theory and praxis, in the same way that religiosity splits into ethics and dogma.

Thus, there would not only be a genesis of technicity, but also a genesis on the basis of technicity, through the splitting of an original technicity into figure and ground [*fond*], the ground corresponding to the functions of totality that are independent of each application of technical gestures, whereas the figure, made of definite and particular schemas, specifies each technique as a manner of acting. The deepest reality [*réalité de fond*] of technics constitutes theoretical knowledge, whereas the particular schemas give us praxis. It is the figural realities of religions which, on the contrary, constitute themselves as a coherent dogma, while the grounding reality [*réalité de fond*] becomes ethical, and detached from dogma; there is an analogy to be found between the praxis that issues forth from technics and the ethics issuing forth from religions, as well as an analogy between the theoretical knowledge of the sciences, which issues forth from technics, and religious 219 dogma; this analogy comes both from the identity of the representative aspect [of both scientific knowledge and religious dogma] or the active aspect [of the praxis of both technics and ethics], and also from the incompatibility that arises from the fact that these different modes of thinking issue forth either from figural realities, or from grounding realities. The purpose [*sens*] of philosophical thought, intervening between the two representative orders and the two active orders of thought, is to make them converge and establish a mediation between them. Now, in order for this mediation to be possible, the very genesis of these forms of thought must be known and accomplished in a complete manner on the basis of previous stages of technicity and religiosity; therefore, philosophical thought must start from the genesis of technicity, integrated into the ensemble of genetic processes that precede it, follow and surround it, not only in order to be able to know technicity in itself, but in order to grasp the problems that dominate the philosophical problematic at their very basis: the theory of knowledge and the theory of action, in their relation with the theory of being.

CHAPTER ONE

The genesis of technicity

I. – The notion of the phase applied to coming-into-being: technicity as a phase. 221

This study postulates that technicity is one of the two fundamental phases of the mode of existence of the whole constituted by man and the world. By phase, we mean not a temporal moment replaced by another, but an aspect that results from a splitting in two of being and in opposition to another aspect; this sense of the word phase is inspired by the notion of a phase ratio in physics; one cannot conceive of a phase except in relation to another or to several other phases; in a system of phases there is a relation of equilibrium and of reciprocal tensions; it is the actual system of all phases taken together that is the complete reality, not each phase in itself; a phase is only a phase in relation to others, from which it distinguishes itself in a manner that is totally independent of the notions of genus and species. The existence of a plurality of phases finally defines the reality of a neutral center of equilibrium in relation to which there is a phase shift. This schema is very different from the dialectical schema, because it implies neither necessary succession, nor the intervention of negativity as a motor of progress; furthermore, opposition, within 222 the schema of phases, only exists in the particular case of a two-phased structure.

The adoption of such a schema founded upon the notion of the phase aims to put into play a principle according to which the temporal development of a living reality proceeds through a split on the basis of an initial, active center, then through a regrouping after the furtherance of each separated reality resulting from this split; each separated reality is the symbol of the other, just as each phase is the symbol of the other phase or phases; no phase, as a phase, is balanced with respect to itself, nor does it contain a complete truth or reality: every phase is abstract and partial, untenable; only the system of phases is in equilibrium in its neutral point; its truth and its reality are this neutral point, the procession and conversion in relation to this neutral point.

We suppose that technicity results from a phase shift of a unique, central, and original mode of being in the world: the magical mode; the phase that balances out technicity is the religious mode of being. Aesthetic thought appears at the neutral point, between technics and religion, at the moment of the splitting of the primitive magical unity: it is not a phase, but rather a permanent reminder of the rupture of unity of the magical mode of being, as well as a reminder of the search for its future unity.

Each phase in turn splits into a theoretical mode and a practical mode; there is thus a practical mode of technics and a practical mode of religion, as well as a theoretical mode of technics and a theoretical mode of religion.

In the same way as the distance between technics and religion gives rise to aesthetic thought, the distance between the two theoretical modes (the technical one and the religious one) gives rise to scientific knowledge, as a mediation between technics and religion. The distance between the practical technical mode and the practical religious mode gives rise to ethical thinking. Aesthetic thought is thus a more primitive mediation between technics and religion than science and ethics, 223 since the birth of science and of ethics requires a prior splitting between the theoretical and the practical mode at the heart of technics and of religion. Out of this arises the fact that aesthetic thought is indeed really situated at the neutral point, prolonging the existence of magic, whereas science on the one hand and ethics on the other oppose each other with respect to the neutral point, since there is the same distance between them as there is between the theoretical and practical mode in technics and religion. If science and ethics could converge and reunite, they would coincide within the axis of neutrality of this genetic system, thereby providing a second analog to the magical unity, above and beyond aesthetic thought, which is its first analog, and which is incomplete since it allows for the phase shift of technics and religion to subsist. This second analog would be complete; it would at once replace magic and aesthetics; but it is perhaps nothing more than a mere tendency playing a normative role, since nothing proves that the distance between the theoretical mode and the practical mode can be completely overcome: this direction defines philosophical research.

In order to indicate the true nature of technical objects, it is thus necessary to resort to a study of the entire genesis of the relations of man and the world; the technicity of objects will then appear as one of two phases of man's relation with the world engendered by the splitting of the primitive magical unity. Must one then consider technicity as a simple moment of genesis? — Yes, in a certain sense, there is indeed something transitory in technicity, which itself splits into theory and praxis and participates in the subsequent genesis of practical and theoretical thought. But in another sense, there is something definitive in the opposition of technicity to religiosity, for one can think that man's primitive way of being in the world (magic) can inexhaustibly furnish an indefinite number of successive contributions capable of splitting into a technical phase and a religious phase; in this way, even though there is effectively a succession in genesis, the successive stages of different geneses are simultaneous within culture, and there exist relations and interactions not only between simultaneous phases, but also between successive stages; not only can technics encounter religion and aesthetic thought, but also science and ethics. Now, if one adopts the genetic postulate, one notices that a science or an ethics can never encounter a religion or a technics on a truly common ground, since the modes of thought are at different levels (for example a science and a technics) and exist at the same time, and they neither constitute a single genetic lineage nor arise from the same sudden outpouring of the primitive magical universe. True and balanced relations only exist between phases of the same level (for example between a technical ensemble and a religion) or between successive degrees of genesis that are part of the same lineage (for example between the stage of technics and religions in the seventeenth century and the contemporary stages of science and ethics). True relations only exist in a genetic ensemble balanced around a neutral point, envisioned in its totality. 224

This is precisely the goal to be attained: the mission of reflexive thought is to lift upright and perfect the successive waves of genesis through which the primitive unity of man's relation with the world splits in two and comes to sustain both science and ethics through technics and religion, between which aesthetic thought develops. In these successive splits, the primitive unity would be lost if science and ethics could not come back together at the end of genesis; philosophical thinking inserts itself between theoretical thought and practical thought by way of an extension of both aesthetic thought and the original magical unity.

Now, in order for the unity of scientific knowledge and ethics to be possible in philosophical thought, the sources of science and ethics must be at the same level, contemporary to each other, and have arrived at the same point of genetic 225 development. The genesis of technics and of religion conditions that of science and of ethics. Philosophy is itself its own condition, for as soon as reflexive thinking has begun, it has the power to perfect whichever of the geneses that has not fully accomplished itself by becoming aware of the sense [*sens*] of the genetic process itself. Hence, in order to be able to pose the philosophical problem of the relations between knowledge and ethics in a profound way, one would first have to complete the genesis of technics and the genesis of religious thought, or at the very least (for this task would be infinite) to know the real direction [*sens*] of these two geneses.

II. – The phase shift from the primitive magical unity

It is therefore necessary to begin with the primitive magical unity of the relations of man and the world in order to understand the true relation of technics to the other functions of human thought; it is through this examination that it is possible to grasp why philosophical thought must realize the integration of the reality of technics into culture, which is possible only by revealing the sense of the genesis of technics, through the foundation of a technology; it is only then that the disparity between technics and religion will be attenuated, which is detrimental to the intention of a reflexive synthesis of knowledge and ethics. Philosophy must found technology, which is the ecumenism of technics, for the sciences and ethics to be able to meet in reflection, a unity of technics and a unity of religious thought must precede the splitting of each of these forms of thought into a theoretical mode and a practical mode.

226   The genesis of a particular phase can be described in itself; but it cannot really be known along with its sense, and consequently grasped in its postulation of unity, unless it is placed back into the totality of the genesis, as a phase in relation with other phases. This is why it is insufficient, for understanding technics, to start from constituted technical objects; objects appear at a certain moment, but technicity precedes them and goes beyond them; technical objects result from an objectivation of technicity; they are produced by it, but technicity does not exhaust itself in the objects and is not entirely contained within them.

If we eliminate the idea of a dialectical relation between successive stages of the relation of man and the world, then what could be the motor of the successive splits in the course of which technicity appears? It is possible to appeal to Gestalt theory, and to generalize the relation it establishes between figure and ground. Gestalt theory derives its basic principle from the hylomorphic schema of ancient philosophy, supported by modern considerations of physical morphogenesis: the structuration of a system would depend on spontaneous modifications tending toward a state of stable equilibrium. However, in reality it seems that it would be necessary to distinguish between a stable equilibrium and a metastable equilibrium. The emergence of the distinction between figure and ground is indeed the result of a state of tension, of the incompatibility of the system with itself, from what one could call the oversaturation of the system; but structuration is not the discovery of the lowest level of equilibrium: stable equilibrium, in which all potential would be actualized, would correspond to the death of any possibility of further transformation; whereas living systems, those which precisely manifest the greatest spontaneity of organization, are systems of metastable equilibrium; the discovery of a structure is indeed at the very least a provisional resolution of incompatibilities, but it is not the destruction of potentials; the system continues   227 to live and evolve; it is not degraded by the emergence of structure; it remains under tension and capable of modifying itself.

If one agrees to accept this corrective and replaces the notion of stability with that of metastability, it seems that Gestalt theory can account for the fundamental stages of the coming-into-being of the relation between man and the world.

Primitive magical unity is the relation of the vital connection between man and the world, defining a universe that is at once subjective and objective prior to any distinction between the object and the subject, and consequently prior to any appearance of the separate object. One can conceive of the primitive mode of man's relation to the world as prior not only to the objectivation of the world, but even to the segregation of objective units in the field that will be the objective field. Man finds himself linked to a universe experienced as a milieu. The emergence of the object only occurs through the isolation and fragmentation of the mediation between man and the world; and, according to the posited principle, this objectivation of a mediation must have as correlative, with respect to the primitive neutral center, the subjectivation of a mediation; the mediation between man and the world is objectivized as technical object just as it is subjectivized as religious mediator; but this objectivation and subjectivation, which are opposition and complementarity, are preceded by an initial relation to the world, the magical stage, in which the mediation is not yet either subjectivized or objectivized, nor fragmented or universalized, and is only the simplest and most fundamental of structurations of the milieu of a living being: the birth of a network of privileged points of exchange between the being and the milieu.

The magical universe is already structured, but according to a mode prior to the segregation of object and subject; this primitive mode of structuration is one that distinguishes figure and ground by marking key-points in the universe. If the universe were devoid of all structure, then the relation between the living being and its milieu could take place in a continuous time and a continuous space, without any privileged moment or place. In fact, preceding the segregation of units, a reticulation of space and time that highlights privileged places and moments institutes itself, as if all of man's power to act and all the world's ability to influence man were concentrated in these places and in these moments. These places and these moments keep hold of, concentrate, and express the forces contained in the ground [*fond*] of reality that supports them. These points and these moments are not separate realities; they draw their force from the ground they dominate; but they localize and focalize the attitude of the living vis-à-vis its milieu.

According to this general genetic hypothesis, we suppose that the primitive mode of existence of man in the world corresponds to a primitive union, prior to any split, of subjectivity and objectivity. The first structuration, corresponding to the appearance of a figure and a ground in this mode of existence, is the one that gives rise to the magical universe. The magical universe is structured according to the most primitive and meaningful of organizations: that of the reticulation of the world into privileged places and privileged moments. A privileged place, a place that has a power, is one that drains from within itself all the force and efficacy of the domain it delimits; it summarizes and contains the force of a compact mass of reality; it summarizes and governs it, as a highland governs and dominates a lowland; the elevated peak is the lord of the mountain,[1] just as the most impenetrable part of the woods is where all its reality resides. The magical world is thus made of a network of places and of things that have a power and that are bound to other things and other places that also have a power. This path, this enclosure, this τέμενος [*temenos*] contains all the force of the land, the key-point of the reality and the spontaneity of things, as well as their availability.

In such a network of key-points, of high-places, there is a primitive lack of distinction between human reality and the reality of the objective world. These key-points are real and objective, but they are that by which the human being is immediately bound to the world, both in order to be influenced by it and in order to act upon it; they are points of contact and of mutual, mixed reality, places of exchange and of communication because they are formed from a knot between the two realities.

---

1. Not metaphorically, but really: it is toward it that the geological folding orients itself and the push that has edified the entire high plateau. The promontory is the firmest part of the chain eroded by the sea.

And magical thought comes first, since it corresponds to the simplest and most concrete, the most vast and flexible structuration: that of reticulation. Within the totality constituted by man and the world a network of privileged points actualizing the insertion of human effort appears as an initial structure, and through which the exchanges between man and the world take place. Each singular point concentrates within itself the capacity to command a part of the world that it specifically represents and whose reality it translates, in communication with man. One could call these singular points the *key-points* commanding over the man-world relation, in a reversible way, for the world influences man just as man influences the world. Such are the peaks of the mountains or certain, naturally magical, mountain passes, because they govern a land. The heart of the forest, the center of a plain are not only metaphorically or geometrically designated realities: they are realities that concentrate the natural powers as they focalize human effort: they are the figural structures in relation to the mass that supports them and constitutes their ground.

When seeking to identify the remnants of magical thought in the context of   230
the current conditions of life, we usually look at superstition as an example of the schemas of magical thought. Superstitions are, in fact, merely a degraded vestige of magical thought, and can only mislead the search for its true essence. One ought, on the contrary, to refer to high, noble, and sacred forms of thought, requiring a fully enlightened effort in order to understand the sense of magical thought. Such is, for example, the affective, representative and voluntary foundation that supports an ascent or an exploration. The desire for conquest and a sense of competition are perhaps a part of the motivation that enables one to go from common existence to these exceptional acts; but what is mostly at stake, when one invokes the desire for conquest, is to legitimize an individual act for a community. In fact, the thought at work in the individual or the small group of those who realize an exceptional act is much more primitive and far richer.

The ascent, the exploration, and more generally any pioneering gesture, consists in connecting with the key-points that nature presents. To climb a slope in order to go toward the summit, is to make one's way toward the privileged place that commands the entire mountain chain, not in order to dominate or possess it, but in order to exchange a relationship of friendship with it. Man and nature are not strictly speaking enemies before this connection at this key-point, but are simply strangers to each other. For as long as it hasn't been climbed, the summit is merely a summit, a place that is higher than the others. The ascent gives it the character of a place that is richer and fuller, and not abstract, a place through which this exchange between the world and man comes to pass. The summit is the place from which the entire mountain chain is seen in an absolute manner, whereas all the sights from all the other places are relative and incomplete, arousing the desire for the view from 231 the summit. An expedition or a navigation allowing one to reach a continent by a definite route do not conquer anything; and yet they are valid according to magical thought, because they allow one to make contact with this continent in a privileged place that is a key-point. The magical universe is made of a network of access points to each domain of reality: thresholds, summits, limits, and crossing points, attached to one another through their singularity and their exceptional character.

This network of limits is not only spatial, but also temporal; there are remarkable dates, privileged moments to begin this or that action. Moreover, the very notion of a beginning is magical, even if all particular value is denied to the date of the beginning; the beginning of an action that is meant to last, the first act in a long series of actions, would not in themselves have any majesty or any particular power of direction, if they weren't considered as governing the duration of the action as well as the rest of the successful or unsuccessful efforts; dates are privileged points in time allowing an exchange between human intention and the spontaneous unfolding of events. Man's insertion into natural coming-into-being is carried out by way of these temporal structures, just as the influence of natural time is exerted on every human life as it becomes a destined end.

In current civilized life, vast institutions are concerned with magical life, but they are hidden by way of utilitarian concepts that justify them indirectly; in particular official holidays, celebrations, and vacations which compensate, with their magical charge, for the loss of magical power that civilized urban life imposes on us. Thus, holiday trips or vacations, which are considered ways for procuring rest and distraction, are in fact a search for old or new key-points; these points can be the big city for the country-dweller, or the countryside for the urbanite, but it is more generally not just any point of the city or countryside; it is the shore or the 232 high mountain, or else the border one crosses in order to arrive into a foreign land. The dates of public holidays are relative to privileged moments in time; sometimes, there can be an encounter between the singular moments and the singular points.

Everyday time and space, in turn, serve as the ground to these figures; dissociated from the ground, the figures would lose their signification; holidays and celebrations are not simply a time of rest with respect to current life, through a halting of current life, but rather a search for the privileged places and dates in relation to the continuous ground.

The figural structure, in primitive magical thought, is inherent to the world, it is not detached from it; it is the reticulation of the universe into privileged key-points through which the exchanges between the living thing and its milieu come to pass. Now, it is precisely this reticular structure that is phase-shifted when one passes from the original magical unity to technics or religion: figure and ground separate by detaching themselves from the universe to which they adhered; the key-points objectivize themselves and only retain their functional characteristics of mediation, they become instrumental, mobile, capable of efficacy in any place and in any moment whatsoever: as a figure, the key-points, detached from the ground whose key they were, become technical objects, transportable and abstracted from the milieu. At the same time, the key-points lose their mutual reticulation and their power of influence from a distance on the reality that surrounded them; as technical objects they have action only through contact, point by point, instant by instant. This rupture of the network of key-points frees the characteristics of ground which, in their turn, detach themselves from their own narrowly qualitative and concrete ground in order to hover over the whole universe, the entirety of space and throughout all of duration, in the form of detached powers and forces, above the world. While the key-points objectivize themselves in the form of concretized tools and instruments, the ground powers subjectivize themselves by personifying 233 themselves in the form of the divine and the sacred (God, heroes, priests).

The primitive reticulation of the magical world is thus the source of opposing objectivation and subjectivation; at the moment of rupture of the initial structuration, the fact that the figure detaches itself from the ground is translated by another detachment: figure and ground detach themselves from their concrete adherence to the universe and follow opposite paths; the figure fragments itself, while the qualities and forces of the ground universalize themselves: this parceling out and this universalization are, for the figure, ways of becoming an abstract figure, and for the ground, a unique abstract ground. This phase shift of mediation into figural characteristics and characteristics of ground translates the appearance of a distance between man and the world; the mediation itself, rather than being a simple structuration of the universe, takes on a certain density; it objectivizes itself in technics and subjectivizes itself in religion, leading to the appearance of the first object in the technical object and of the first subject in divinity, when there was hitherto only a unity of the living and its milieu: objectivity and subjectivity appear between the living and its milieu, between man and the world, at a moment when the world does not yet have a complete status of object nor man a complete status of subject. One can furthermore note that objectivity is never completely coextensive with the world, any more than subjectivity is completely coextensive with man; it is only when the world is viewed from a technicist perspective and man from a religious perspective that it appears that one can be said to be entirely object, and the second entirely subject. Pure objectivity and pure subjectivity are modes of mediation between man and the world, in their initial form.

234 Technics and religion are the organization of two symmetrical and opposed mediations; but they form a couple, because they are each only a phase of the primitive mediation. In this sense, they possess no definitive autonomy. What is more, even taken in the system they form, they cannot be considered as enclosing all of the real, since they are between man and the world, but do not contain all of the reality of man and world, and cannot apply to it in a complete way. Directed by the gap that exists between these two opposite aspects of mediation, science and ethics deepen the relation between man and the world. With respect to science and ethics, the two primitive mediations play a normative role: science and ethics are born in the space defined by the gap between technics and religion, following a median direction; the direction exercised by the precedence of technics and of religion before science and ethics is of the same order as that exerted by the lines limiting an angle on the bisector of that angle: the sides of the angle can be indicated by short segments, while the bisector can be indefinitely extended; in the same way, on the basis of the gap between very primitive technics and religion, a very elaborate science and ethics can progressively be constructed, that is guided rather than limited by the basic conditions of technics and religion.

The origin of the split that has given rise to technical thought and religious thought can be attributed to a truly functional primitive structure of reticulation. This split has separated figure and ground, the figure giving the content of technics, and the ground that of religion. While, in the magical reticulation of the world, figure and ground are reciprocal realities, technics and religion appear when figure and ground detach themselves from one another, thereby becoming mobile, fragmentable, displaceable and directly manipulable because they are not bound to the world. Technical thought retains only the schematism of structures, of that 235 which makes up the efficacy of action on the singular points; these singular points, detached from the world whose figure they were, also detached from one another, losing their immobilizing reticular concatenation, become capable of being fragmented and available, as well as reproducible and constructible. The elevated place becomes an observation post, a watchtower built on the plain, or a tower placed at the entrance of a gorge. Often, a nascent technics need go no further than modifying a privileged place, as when constructing a tower on the summit of a hill, or by placing a lighthouse on a promontory, at the most visible point. But technics can also completely create the functionality of privileged points. It merely retains the figural power of the natural realities, not the placement and natural localiza-tion on a ground that is determined and given prior to any human intervention. Fragmenting the schematisms more and more, it turns the thing into a tool or an instrument, in other words a detached fragment of the world, capable of operating efficiently and in any place and under any conditions, point by point, according to the intention directing it and the moment man wants it. The availability of the technical thing consists in being liberated from the enslavement to the ground of the world. Technics is analytical, operating progressively and through contact, setting aside the liaison through influence. In magic, the singular place enables action on a domain in its entirety, just as it suffices to speak to the king in order to win over an entire people. On the contrary, in technics the whole of reality must be traversed, touched, and treated by the technical object, detached from the world and applicable to any point and at any moment. The technical object distinguishes itself from the natural being in the sense that it is not part of the world. It intervenes as mediator between man and the world; it is, in this respect, the first detached object, since the world is a unity, a milieu rather than an ensem-ble of objects; there are in fact three types of reality: the world, the subject, and the object, which is intermediary between the world and the subject, whose initial   236 form is that of the technical object.[2]

III. – The divergence of technical thought and of religious thought

Technical thought — resulting from the rupture in the primitive structure of the magical world's reticulation, and retaining the figural elements that can be depos-ited in objects, tools, or instruments — gains an availability from this detachment that enables it to apply itself to every element of the world. However, this rupture also produces a deficit: the technical tool or instrument has only retained figural characteristics detached from the ground to which they were once directly attached since they arose out of an initial structuration that provoked the emergence of figure and ground within a reality that had been one and continuous. In the mag-ical universe, the figure was the figure of a ground and the ground, ground of a figure; the real, the unity of the real, was at once figure and ground; the question of a possible lack of the figure's efficacy on the ground or of the ground's influ-ence on the figure could not arise, since ground and figure merely constituted a

---

2.   Variation in the proofs: "there are in fact three types of reality: the world, man, and the object, intermediary between the world and man, whose first form is the technical object." — Ed.

single unity of being. Conversely, in technics, after the rupture, what the technical object retained and maintained of figural characteristics will henceforth encounter any ground whatsoever, an anonymous, foreign ground. The technical object has become a bearer of form, a remnant of figural characteristics, and it seeks to apply this form to a ground that is now detached from the figure, having lost its intimate relation of belonging, and capable of being informed by whichever form it encounters, but in a violent, more or less imperfect manner; figure and ground have become foreign and abstract in relation to each other.

    The hylomorphic schema doesn't describe only the genesis of living beings; perhaps it doesn't even essentially describe it. Perhaps it does not even come from the reflected and conceptualized experience of technics: before the knowledge of the living being and before the reflection on technics, there is this implicit adequation of figure and ground, ruptured by technics; if the hylomorphic schema appears to emerge from technical experience, it is as a norm or as an ideal rather than as an experience of the real; technical experience, putting into play vestiges of figural elements and vestiges of the ground characteristics gives new life to the primary intuition of a mutual belonging of matter and form, of a coupling preceding all splitting. In this sense, the hylomorphic schema is true, not because of the logical use that has been made of it in ancient philosophy, but as an intuition of a structure of the universe for man prior to the birth of technics. This relation cannot be hierarchized, there cannot be more and more abstract successive stages of matter and form, since the real model of the relation of matter and form is the first structuration of the universe into ground and figure; indeed, this structuration can only be true if it is not abstract, if it is on a single level; the ground is really ground and the figure is really figure, it cannot become ground for a higher figure. The manner in which Aristotle describes the relations between form and matter, in particular the supposition that matter aspires to form ("matter aspires toward form as the female to the male"), is already far from primitive magical thought, for this aspiration can exist only if there is a prior detachment; but here, there is just one being, which is both matter and form. Furthermore, perhaps it should not be said that the individual being alone has form and matter; since the appearance of a figure-ground structure is prior to any segregation of units; the mutual relation of correspondence of such a key-point and of such a ground neither presupposes this key-point to be isolated from the network of other key-points nor this ground to be without continuity with the other grounds: a universe is what is structured in this way, and not a set [*ensemble*] of individuals; after the rupture of the primitive reticulation the first detached beings to appear are technical objects and religious

subjects, and they are charged either with figural characteristics or with characteristics of ground: hence they do not fully possess form and matter.

The dissociation of the primitive structuring of the magical universe entails a
series of consequences for technics and religion, and through them, conditions the
subsequent coming-into-being of science and ethics. Unity belongs to the magical
world. The phase shift opposing technics and religion irreducibly leaves the content of technics with a status lower than unity and that of religion with a status
higher than unity. This is where all the other consequences come from. In order
to fully understand the status of the technicity of objects, one must grasp this
coming-into-being that puts the primitive unity out of phase. Religion, retaining
its ground-characteristics (homogeneity, qualitative nature, lack of distinction of
elements within a system of mutual influences, long distance action through space
and time, engendering ubiquity and eternity), represents the putting into play of
these functions of totality. A particular being, a defined object of attention or effort,
is always considered, in religious thought, to be smaller than the real unity, inferior
to the totality and included in it, surpassed by the totality of space and preceded
and followed by the immensity of time. The object, the being, the individual, subject or object, are always grasped as less than unity, dominated by a sensed totality
that infinitely surpasses them. The source of transcendence lies in the function of
totality that dominates the particular being; according to the religious view, this   239
particular being is understood with reference to a totality in which it participates,
on which it exists, but which it can never completely express. Religion universalizes
the function of a totality, which is dissociated and consequently freed of all figural
attachment limiting it; the grounds related to the world in magical thought, and
consequently limited by the very structuration of the magical universe, become
in religious thought a limitless spatial as well as temporal background; they retain
their positive qualities of ground (the forces, the powers, the influences, the quality), but rid themselves of their limits and of their belonging which attached them
to the *hic et nunc*. They become absolute ground, a grounding totality [*totalité de
fond*]. The universe is promoted on the basis of these freed up and, to a certain
extent, abstract magical grounds.

After the disjunction of ground and figure, religious thought preserves the other
part of the magical world: the ground, with its qualities, tensions, and forces; but,
like the figural and technical schemas, this ground itself also becomes something
detached from the world, abstracted from the primitive milieu. And in the same
way that figural schemas of technics, once freed from their adherence to the world,
fix themselves onto the tool or instrument by objectivizing themselves, so too

the ground qualities, made available by the mobilization of figures through technicity, fix themselves onto subjects. The technical objectivation that leads to the emergence of the technical object, mediating between man and the world, has religious subjectivation as its counterpart. In the same way that technical mediation establishes itself by means of a thing that becomes the technical object, religious mediation appears by virtue of the fixing of the characteristics of ground onto real or imaginary subjects, divinities or priests. Religious subjectivation normally leads to mediation through of the priest, while technical mediation leads to the mediation through the technical object. Technicity retains the figural characteristic of the primitive complex of man and the world, while religiosity retains the character of ground.

Technicity and religiosity are not degraded forms of magic, or relics of magic; they come from the splitting in two of the primitive magical complex, the original reticulation of the human milieu, into figure and ground. It is through their coupling, and not in and of themselves, that technics and religion are the heirs of magic. Religion is not more magical than technics; it is the subjective phase of the result of a split, while technics is the objective phase of this same split. Technics and religion are contemporaries of one another and, considered on their own, they are impoverished with respect to the magic from which they come.

Religion thus has by nature the vocation to represent the demand for totality; when it splits into a theoretical mode and into a practical mode, it becomes by way of theology the demand for a systematic representation of the real, according to an absolute unity; through morality, it becomes the demand, from the ethical point of view, for absolute norms of action, justified in the name of totality, superior to any hypothetical, i.e., particular imperative; to both science and ethics it brings a principle of reference to totality, which is the aspiration to the unity of theoretical knowledge and to the absolute character of the moral imperative. The religious inspiration constitutes a permanent reminder of the relativity of a particular being with respect to an unconditional totality, going beyond all objects and subjects of knowledge and of action.

Conversely, technics receive content that is always below the status of unity, because the schemas of efficacy and the structures that result from the fragmentation of the primitive network of key-points cannot apply to the totality of the world. Technical objects are multiple and fragmentary by nature; technical thought, enclosed within this plurality, can progress, but only by multiplying technical objects, without being able to recapture the primitive unity. Even by infinitely multiplying technical objects, it is impossible to recover an absolute adequation with the world, because each of the objects attacks the world only in a single point and at a single moment; it is localized, particularized; by adding technical objects to one another, one can neither recreate a world, nor recover the contact with the world in its unity, which was the goal of magical thought.

In its relation to a determined object or to a determined task, technical thought is always at a level inferior to that of unity: it can present several objects, several means, and choose the best; but it nevertheless always remains inadequate to the whole of the unity of the object or of the task; each schema, each technical object, each technical operation is dominated and guided by the whole from which it derives its means and its orientation, and which provides it with a never attained principle of unity that it translates by combining and multiplying its schemas.

The vocation of technical thought is by nature representing the point of view of the element; it adheres to the elementary function. Technicity, by introducing itself into a domain, fragments it and leads to the appearance of a chain of successive and elementary mediations, governed by the unity of the domain and subordinated to it. Technical thought conceives of an overall functioning [*fonctionnement d'en-semble*] as a series of elementary processes, acting point by point and step by step; it localizes and multiplies the mediation schemas, always remaining below unity. The element, in technical thought, is more stable, better known, and in a certain way more perfect than the ensemble; it is really an *object*, whereas the ensemble always remains to a certain extent inherent in the world. Religious thought finds the opposite balance: for religious thought, totality is that which is more stable, 242 stronger, and more valid than the element.

Technics bring, as much in the theoretical domain as in the ethical domain, a concern for the element. In the sciences, the contribution of technics has consisted in allowing a representation of phenomena taken one by one according to a decomposition into simple, elementary processes comparable to the operations of technical objects; such is the role of the mechanistic hypothesis that enables Descartes to represent the rainbow as the overall result of the point by point trajectory of each luminous corpuscle in each droplet of water in a cloud; it is according to the same method that Descartes explains the functioning of the heart, decomposing a complete cycle into simple successive operations, and showing that the functioning of the whole is the result of the play of elements necessitated by their particular disposition (for example that of each valve). Descartes doesn't ask himself why the heart is made in this way, with valves and cavities, but how it functions given that this is how it is made. The application of schemas drawn from technics does not account for the existence of the totality, taken in its unity, but only for the point by point and instant by instant functioning of this totality.

In the ethical domain, technical thought not only introduces means of action, which are fragmentary and tied to the capacities of each object becoming a utensil, but also a certain duplication of the action by technicity; a definite human action, considered in its result, could have been accomplished by a determinate technical functioning going through different stages; elements and moments of action have their technical analog; an effort of attention, of memory, could have been replaced by a technical operation; technicity provides a partial equivalence of the results of an action; it accentuates the awareness of the action by the being who accomplishes it in the form of results; it mediates and objectivizes the results of the action by comparison with those of the technical operation, performing a decomposition of the action into partial results, into elementary accomplishments. In the same way that in the sciences, technicity introduces the search for a *how* through the decomposition of an overall phenomenon [*phénomène d'ensemble*] into elementary operations, so too in ethics, technicity introduces the search for a decomposition of a global action into elements of action; the total action being envisaged as that which leads to a result, the decomposition of the action called for by technics considers the elements of an action as movements obtaining partial results. Technicity presupposes that an action is limited to its results; it is not concerned with the subject of the action taken in its real totality, nor even with an action in its totality, insofar as the totality of the action is founded on the unity of the subject. The concern with the result in ethics is the analog of the search for a *how* in the sciences; result and process remain below the unity of action or of the whole [*ensemble*] of the real.

The postulation of an absolute and unconditional justification that religion directs at ethics translates into the search for intention, as opposed to the search for the result that is inspired by technics. In the sciences, religious thought introduces a quest for absolute theoretical unity, rendering necessary a search for the sense of the coming-into-being and of the existence of given phenomena (hence answering a *why?*), while technical thought brings with it an examination of the *how?* of each of the phenomena.

In possessing a content that is at a lower level than unity, technical thought is the paradigm of all inductive thinking, whether in the theoretical order, or in the practical order. It contains this inductive process within itself, prior to any separation into a theoretical mode and a practical mode. Induction, in fact, is not only a logical process, in the strict sense of the term; one can consider as inductive any approach whose content has a lower status than that of unity, and which strives to attain unity, or which at least tends toward unity on the basis of a plurality of elements where each is lower than unity. What induction grasps, what it starts from, is an element that is not in itself sufficient and complete, that does not constitute a unity; it thus goes beyond each particular element, combining it with other elements that are themselves particular in order to attempt to find an analog of unity: within induction there is a search for the ground of reality on the basis of figural elements that are fragments; to want to find the law beneath phenomena, as with the induction of Bacon or J. S. Mill, or to seek only to find what is common to all individuals of a same kind, as with Aristotle's induction, is to postulate that beyond the plurality of phenomena and individuals, a stable and common ground of reality exists, which is the unity of the real.

It is no different for any ethics that would come directly from technics; to want to compose the whole of the duration of life from a series of instants, extracting from each situation what is pleasant in it and to want to construct the happiness of life from the accumulation of these pleasant elements, as is done in ancient Eudemonism or in Utilitarianism, is to proceed in an inductive manner, by trying to replace the unity of life's duration and the unity of human aspiration with a plurality of instants and with the homogeneity of all successive desires. The elaboration to which Epicureanism submits desires has as its sole aim to achieve the incorporation into the continuity of existence by proceeding in a cumulative manner: for this purpose, each of the desires must be dominated by the subject, enveloped in it, smaller than unity, so as to be able to be treated and manipulated as a true element. This is why the passions are eliminated, since they cannot be treated as elements; they are larger than the unity of the subject; they dominate it, come from further afield than it does and tend to advance further than it does, obliging it to go beyond its limits. Lucretius tries to destroy the passions from 245 within, by showing that they are based on errors; he does not, in fact, take into account the element of tendency in passion, in other words, this force that inserts itself into the subject, but that is vaster than it, and in relation to which it appears as a very limited being; tendency cannot be considered as being contained in the subject understood as unity. Wisdom, having reduced the forces at the origin of action to a lower status than that of the unity of the moral subject, can organize them as elements and reconstruct a moral subject within the natural subject; this moral subject, however, never completely reaches the level of unity; between the reconstructed moral subject and the natural subject, there remains a void that is impossible to fill; the inductive approach remains within plurality; it constructs a bundle of elements, but this bundle cannot be equivalent to a real unity. Every ethical technique leaves the moral subject dissatisfied, because they do not grasp its unity; the subject cannot content itself with a life that would be a sequence of happy instances, even an interrupted one; a life that is perfectly successful, element by element, is not yet a moral life; it lacks unity, which is what makes it the life of a subject.

But religious thought, inversely, which is the foundation of obligation, creates a search in ethical thinking for an unconditional justification that makes each act and every subject appear as inferior to real unity; related back to a totality that dilates infinitely, the moral act and subject derive their meaning only from their relation with this totality; the communication between the totality and the subject is precarious, because at every instant the subject is brought back to the dimension of its own unity, which is not that of the totality; the ethical subject is de-centred by the religious requirement.

CHAPTER II

I. – *Technical thought and aesthetic thought*

According to a genetic hypothesis such as this one, it would be best not to con- 247
sider the different modes of thought as parallel to one another; thus, one cannot
compare religious thought and magical thought because they are not on the same
level; but on the contrary it is possible to compare technical thought and religious
thought, because they are contemporary to each other; in order to compare them,
it is not enough to determine their particular characteristics, as if they were the
species of a genus; one must return to the genetic realization of their formation,
for they exist as a couple, resulting from the split in primitive complete thought,
which was magical thought. As for aesthetic thought, it is never characteristic of
a limited field or of a determinate species, but only of a tendency; it is that which
maintains the function of totality. In this sense, it can be compared to magical
thought, provided however that one specifies that it does not contain, as magical
thought does, the possibility of splitting into technics and religion; indeed, far
from going in the direction of a split, aesthetic thought is what maintains the 248
implicit memory of unity; from one of the phases of splitting, it calls upon the
other complementary phase; it seeks totality in thought and aims at recomposing
a unity through an analogical relation where the appearance of phases could create
the mutual isolation of thought in relation to itself.

Such a way of approaching aesthetic effort would undoubtedly be untenable if one thus wanted to characterize works of art such as they exist in their institutional state in a given civilization, and even more so, if one wanted to define the essence of aestheticism. But, in order for works of art to be possible, they must be made possible by a fundamental tendency in the human being, and by the ability to experience the aesthetic impression in certain real and vital circumstances. The artwork that is part of a civilization uses aesthetic feeling and satisfies, sometimes artificially and in an illusory manner, man's tendency to seek a complement with respect to a totality, when he exerts a certain type of thought. It would be insufficient to say that the work of art manifests the nostalgia for magical thought; the work of art, in fact, grants us the equivalent of magical thought, since it recovers — on the basis of a given situation, and according to an analogical structural and qualitative relation — a universalizing continuity with respect to other situations and to other possible realities. The work of art re-establishes a reticular universe at least for perception. But the work of art doesn't really reconstruct the primitive magical universe: this aesthetic universe is partial, integrated, and contained in the real and actual universe that has emerged from the split. In fact, the work of art above all sustains and preserves the ability to experience aesthetic feeling, just as language sustains the ability to think, without nevertheless itself being identical to thought.

Aesthetic feeling is not relative to an artificial work; it signals, in the exercise of a mode of thinking that is subsequent to the split, a perfection of completion that makes the ensemble of acts of thought capable of surpassing the limits of its domain so as to evoke the completion of thought in other domains: a technical work perfect enough to be equivalent to a religious act, a religious work perfect enough to have the organizational and operational force of a technical activity give off a feeling of perfection. Imperfect thought stays within its domain; the perfection of thought allows the μετάβασις εἰς ἄλλο [metábasis eis állo] that gives the fulfillment of a particular act a universal significance through which an equivalent of the magical totality, which had been abandoned at the origin, is recovered at the end of human effort; and the world itself must be present and authorize this achievement after a long detour. The artistic impression implies the feeling of the complete perfection of an act, a perfection that objectively gives it a radiance and an authority through which it becomes a remarkable point of lived reality, a knot of experienced reality. This act becomes an outstanding point of the network of human life integrated within the world; from this outstanding point a higher kinship with others is created, reconstituting an analog of the magical network of the universe.

The aesthetic character of an act or a thing is its function of totality, its existence, both objective and subjective, as an outstanding point. Any act, any thing, any moment has in itself the ability to become an outstanding point of a new reticulation of the universe. Every culture selects the acts and situations that are apt to become outstanding points; but culture is not what creates the aptitude of a situation to become an outstanding point; it only forms a barrage against certain types of situations, leaving narrow straights for aesthetic expression with respect to the spontaneity of the aesthetic impression; culture intervenes as limit rather than as creator.

The destiny of aesthetic thought, or more precisely of the aesthetic inspiration   250 of all thought tending toward its own completion, is to reconstitute, within each mode of thinking, a reticulation that coincides with the reticulation of other modes of thinking: the aesthetic tendency is the ecumenism of thought. In this sense, beyond even the maturity of each of the genera of thought, there occurs a final reticulation that once again brings the separate types of thought which emerged from the shattering of primitive magic closer together. The first stage of each type of thought's development is isolation, non-adherence to the world, abstraction. Then, through its very development, each type of thought, which initially rejected what is not itself and behaved as a species, after having affirmed itself according to the unconditional monism of principles, pluralizes itself and widens according to a principle of plurality; one could say that each thought tends to become reticular and once more to adhere to the world after having distanced itself from it. After having mobilized and detached the schematic figures of the magical world from the world, technics return to the world to ally itself with it through the coinciding cement and rock, of the cable and the valley, the pylon and the hill; a new reticulation establishes itself, chosen by technics, privileging certain places of the world, in a synergetic alliance of technical schemas and natural powers. This is where aesthetic feeling appears, in this agreement and this surpassing of technics once more becoming concrete, integrated, and attached to the world through the most outstanding key-points. The mediation between man and the world becomes itself a world, the structure of the world. In the same manner, religious mediation accepts concretizing itself, after a dogmatism that was detached from the concreteness of the universe and having mobilized every dogma to conquer every representative of humankind, in other words religious mediation accepts attaching itself to each culture and to each human group according to relatively pluralistic modalities; unity becomes the unity of a network rather than being a monist unity of a single principle and a single faith. The maturity of technics and of religions tends toward   251 re-incorporation into the world, the geographical world for technics, the human world for religions.

To this day, it does not appear possible for the two reticulations, that of technics within the geographical world and that of religions in the human world, to analogically encounter each other in a real, symbolic relation. And yet only in this way could the aesthetic impression state the rediscovery of the magical totality, by indicating that the forces of thought have once again found one another. Aesthetic feeling, common to both religious thought and technical thought, is the only bridge that could allow for the linking of these two halves of thought that result from the abandonment of magical thought.

Philosophical thought, in order to know how to deal with the contribution of technics and religion at the level of the distinction between the theoretical and practical modalities, can thus ask itself how aesthetic activity deals with this contribution at the level prior to the distinction of these modalities. What is broken in the move from magic to technics and religion, is the first structure of the universe, in other words the reticulation of key-points, which is the direct mediation between man and the world. And aesthetic activity preserves precisely this structure of reticulation. It cannot really preserve it in the world, since it cannot substitute itself for technics and religion, which would be to recreate magic. But it preserves it by constructing a world in which it can continue to exist, and which is at once technical and religious; it is technical because it is constructed rather than natural, and because it uses the power of applying technical objects to the natural world in order to make the world of art; it is religious in the sense that this world incorporates the forces, the qualities, the characteristics of ground that technics leave out; instead of subjectivating them as religious thought does by universalizing them, instead of objectivating them by enclosing them in the tool or instrument, as technical thought does when it works on the basis of dissociated figural structures, aesthetic thought limits itself to concretizing the ground qualities via technical structures, staying in the space between religious subjectivation and technical objectivation: it thus makes the aesthetic reality, which is a new mediation between man and the world, an intermediate world between man and the world.

Aesthetic reality in fact cannot be said to be either properly object or properly subject; there is, of course, a relative objectivity to the elements of this reality; but aesthetic reality is not detached from man and from the world like a technical object; it is neither tool nor instrument; it can stay attached to the world, for instance by being an intentional organization of a natural reality; it can also stay attached to man, by becoming a modulation of the voice, a turn of phrase, a way of dressing; it does not have this necessarily detachable character of the instrument; it can remain integrated, and normally it does stay integrated within human reality or the world; a statue is not placed just anywhere, a tree is not planted just anywhere. There is a beauty of things and of beings, and a beauty in the ways of being, and aesthetic activity starts by experiencing it and by organizing it, by respecting it when it is naturally produced. Conversely, technical activity constructs separately, detaching its objects, and applying them to the world in an abstract and violent way; even when the aesthetic object is produced in a detached way, as a statue or a lyre, this object remains a key-point of a part of the world and of human reality; the statue thus placed before a temple is what makes sense for a defined social group, and the mere fact that it is placed, in other words that it occupies a key-point that it uses and reinforces but does not create, shows that it is not a detached object. One can say that a lyre is an aesthetic object, insofar as it produces sounds, but 253 the sounds of the lyre are aesthetic objects only to the extent that they concretize a certain mode of expression, of communication, that already exists in man; the lyre can be carried like a tool but the sounds it produces and which constitute the true aesthetic reality are integrated into human reality and the reality of the world; the lyre can only be listened to in silence or with certain determinate sounds like that of the wind or the sea, and not with the noise of the voice or the murmur of a crowd; the sound of the lyre must integrate itself into the world, in the same way the statue becomes integrated. Conversely, the technical object, insofar as it is a tool, does not become integrated because it can act and function anywhere.

It is indeed this integration that defines the aesthetic object, and not imitation: a piece of music that imitates noise cannot become integrated into the world, because it replaces certain elements of the universe (for instance the noise of the sea) rather than completing them. A statue, in a certain sense, imitates a man and replaces him, but this is not why it is an aesthetic work; it is an aesthetic work because it becomes integrated into the architecture of a town, marks the highest point of a promontory, forms the endpoint of a wall, or sits atop a tower. Aesthetic perception senses a certain number of requirements: there are empty spaces that need filling, rocks that need to bear a tower. There are a certain number of outstanding places in the world, exceptional points that attract and stimulate aesthetic creation, as there are a certain number of particular, radiant moments in a human life, that distinguish themselves from others, that call for a work of art. The work, resulting from this requirement of creation, from this sensitivity to places and moments of exception, does not copy the world or man, but rather extends them and becomes integrated with them. Even if it is detached, the aesthetic work does not arise from a rupture in the universe or in the life time of man; it comes as a surplus of already given reality, bringing it constructed structures, but constructed on foundations that are a part of the real and which become integrated into the world. The aesthetic work thus makes the universe bud, extending it by establishing a network of works, in other words by establishing radiating realities of exception, key-points of a universe that is at once human and natural. More detached from the world and from man than the magical universe's old network of key-points, the spatial and temporal network of artworks is a mediation between man and the world which preserves the structure of the magical world.

It would, undoubtedly, be possible to affirm that there is a continuous transition between the technical and the aesthetic object, since there are technical objects that have an aesthetic value and that can be said to be beautiful: the aesthetic object can then be conceived as not being integrated into a universe, and thus like the technical object, can be considered as detached, since a technical object can be considered as an aesthetic object.

In fact, technical objects are not inherently beautiful in themselves, unless one is seeking a type of presentation that answers directly to aesthetic concerns; in this case, there is a true distance between the technical object and the aesthetic object; it is as if there were in fact two objects, the aesthetic object enveloping and masking the technical object; this is the case for instance when one sees a water tower, built near a feudal ruin, camouflaged by added crenels and painted the same color as the old stone: the technical object is contained in this fake tower, with its concrete tank, its pumps, its tubes: the hoax is silly, and seen as such from the very first glance; the technical object retains its technicity beneath its aesthetic cover, hence the conflict that arises which gives the impression of the grotesque. Every disguise of a technical object generally produces the uncomfortable impression of a fake, and appears like a materialized lie.

But in certain cases there is a beauty proper to technical objects. This beauty appears when these objects become integrated within a world, whether it be geographical or human: aesthetic feeling is then relative to this integration; it is like a gesture. The sails of a ship are not beautiful when they are at rest, but when the wind billows and inclines the entire mast, carrying the ship on the sea; it is the sail in the wind and on the sea that is beautiful, like the statue on the promontory. The lighthouse by the reef dominating the sea is beautiful, because it is integrated as a key-point of the geographical and human world. A line of pylons supporting the cables that traverse a valley is beautiful, whereas the pylons, seen on the trucks that bring them, or the cables, on the big rolls that serve to transport them, are neutral.

A tractor, in a garage, is merely a technical object; however, when it is at work plowing, leaning into the furrow while the soil is turned over, it can be perceived as beautiful. Any technical object, mobile or fixed, can have its aesthetic epiphany, insofar as it extends the world and becomes integrated into it. But it is not only the technical object that is beautiful: it is the singular point of the world that the technical object concretizes. It is not only the line of pylons that is beautiful, it is the coupling of the lines, the rocks, and the valley, it is the tension and flexion of the cables: herein resides a mute, silent and ever continued operation of technicity applying itself to the world.

The technical object is not beautiful in every circumstance; it is beautiful when it encounters a singular and remarkable place in the world; the high voltage line is beautiful when it traverses a valley, the car when it turns, the train when it enters or exits a tunnel. The technical object is beautiful when it has encountered a ground that suits it, whose own figure it can be, in other words when it completes and expresses the world. The technical object can even be beautiful with respect to an object that is larger than itself serving as its ground, in some ways as its universe. The radar antenna is beautiful when it is seen from the point of view of a ship, sitting atop the highest super-structure; placed on the ground, it is nothing more 256 than a rather crude cone, mounted on a pivot; it was beautiful as the structural and functional completion of this whole [*ensemble*] that is the ship, but it is not beautiful in itself and without reference to a universe.

This is why the discovery of the beauty of technical objects cannot be left to perception alone: the function of the object needs to be understood and thought; in other words, a technical education is needed if the beauty of technical objects is to appear as an integration of technical schemas into a universe, within the key-points of this universe. How, for instance, could the beauty of a radio relay placed on a mountain, and oriented toward another mountain where another relay is placed, appear to the one who only sees a tower of mediocre height, with a parabolic grid in which a very small dipole is placed? All of these figural structures need to be understood as emitting and receiving the bundle of directed waves that propagates from one tower to another, through the clouds and the fog; it is with respect to this invisible, imperceptible, and real, actual transmission that the whole [*ensemble*] formed by the mountains and the towers is beautiful, for the towers are placed at the key-points of the two mountains in order to constitute the wireless cable; this type of beauty is as abstract as that of a geometric construction, and the function of the object needs to be understood in order for its structure, and the relation of this structure with the world, to be correctly imagined, and aesthetically felt.

The technical object can be beautiful in a different way, through its integration into the human world that it extends; thus a tool can be beautiful in action when it properly adapts itself so well to a body that it somehow seems to be a natural extension of it and whose structural characteristics it appears to amplify; a dagger is only beautiful in the hand that holds it; a tool, a machine or a technical ensemble, are equally beautiful when they become integrated within the human world and cover it over in expressing it; if the alignment of boards in a telephone center is beautiful, then it is not beautiful in itself or in its relation with the geographical world, since it can be anywhere; it is beautiful because these luminous flashes that trace the multi-colored and moving constellations represent instant by instant the real gestures of a multitude of humans, attached to one another through the crossing of these circuits. The telephone call center is beautiful in action, because at every instant it is the expression and realization of an aspect of the life of a city and of a region; a light is someone waiting, an intention, a desire, imminent news, a ringing telephone that one won't hear but that will resound far away in another house. Here we witness the beauty found within the action; it is not simply instantaneous, but is also made up of the rhythms of use in peak hours and evening hours. The telephone call center is beautiful not because of its characteristics as an object, but because it is a key-point in collective and individual life. In the same vein, a traffic light [*sémaphore*] on a train platform is not beautiful in itself, but is beautiful by way of its functioning as a traffic light, which is to say through its power to indicate, to signify a stop or a track to be left free. In the same way again the Hertzian modulation we receive, as a technical reality, from a different continent, barely audible, made momentarily unintelligible underneath the static and distortions, is technically beautiful, because it arrives charged with the overcoming of obstacles and distance, bringing us the testimony of a faraway human presence, whose sole epiphany it is. Hearing a nearby powerful transmitter is not technically beautiful, because its value is not transformed by this power to reveal man, to manifest an existence. And it is not only the overcoming of difficulty that makes the reception of a signal emanating from a different continent beautiful; it is the power that this signal has for making a human reality emerge for us, which it extends and manifests in actual existence, by rendering it perceptible for us, when it would have otherwise remained unknown despite being contemporary with ours. "White noise" has as much technical beauty as a meaningful signal, when it bears within itself witness to a human being's intention to communicate; the reception of background noise or of a simple continuous sinusoidal modulation can be technically beautiful when it becomes integrated into a human world.

One can thus say that the aesthetic object is not strictly speaking an object, but rather the extension of the natural or human world that remains integrated within the reality that bears it; it is an outstanding point in a universe; this point is the result of an elaboration and benefits from technicity; but it is not arbitrarily placed in the world; it represents the world and focalizes its ground forces and qualities, like a religious mediator; it keeps itself in an intermediary state between pure objectivity and subjectivity. When the technical object is beautiful, it is because it has been integrated into the natural or human world, just like aesthetic reality.

Aesthetic reality distinguishes itself from religious reality in that it can neither be universalized nor subjectivized; the artist is not confused with the work, and, where a certain idolatry arises, it is recognized as idolatry; it is the work of art's technicity that prevents aesthetic reality from being confused with the universal function of totality; the work of art remains artificial and localized, produced at a certain moment; it is not anterior and superior to the world and to man. The set [*ensemble*] of all works of art perpetuates the magical universe and maintains its structure: it marks the neutral point between technics and religion.

The aesthetic universe, however, is far from being a residue, a simple remnant of a previous epoch; it represents the meaning of coming-into-being diverging in the passage from magic to technics and religion, but eventually having to re-converge toward unity; the immanence of an aesthetic concern, within technics as within religion, is the sign that technical thought, and indeed religious thought as well, represent only a phase of complete thought. Technics and religion cannot communicate directly, but they can communicate through the mediation of aesthetic activity; a technical object can be beautiful in the same way a religious gesture can be beautiful, when they are integrated within the world at an outstanding place and time. A norm of beauty exists within these two opposite modes of thinking, a norm that makes them tend toward one another by applying them to the same universe. Through the aesthetic work of art, the religious act is integrated, because it is the religious act itself that becomes the work; a chant, a canticle, or a celebration are integrated to the *hic et nunc*. The religious gesture is beautiful when it acts as an extension of the natural and the human world. A sacrament is thus a religious gesture, and it is beautiful when it becomes integrated into the world, in a certain place and a certain moment, because it is applied toward determinate people: the qualities of ground once more encounter structures; it is through the beauty of the celebration that religious thought rediscovers a network of moments and places that have a religious value; religious gestures are beautiful when they belong to a time and an era, not because of their external ornaments that have no connection

259

with the world; these ornaments that have no connection to a time or a place isolate religious thought within the vain sterility of a ritual; they are of the order of the grotesque, like the technical object concealed by an aesthetic mask. Religious thought is beautiful when it integrates the function of totality into a spatio-temporal network, when it intervenes with the forces and qualities of the ground of the entire universe in a place and a moment. Furthermore, as in technical thought, this aesthetically valid reintegration can only occur if it meets with the key-points of the natural or human world. A temple, or a sanctuary, are not built by chance, in an abstract way, without relation with the world; there are places of the natural world that call for a sanctuary, just as there are moments in human life that call for a sacramental celebration. In order for the aesthetic impression to arise in religious thought, religion must be established in its own right, containing the forces and qualities of the universe's ground; but also the natural and the human world must be waiting to extend and concretize themselves in religious places and moments according to a norm that is, in a deep way, aesthetic.

Aesthetic reality is thus a surplus to given reality, but according to lines that already exist in given reality; it is what reintroduces the figural functions and the functions of ground into given reality which, in the moment of the magical universe's dissociation, had become technics and religion. Without aesthetic activity, there would only be a neutral zone between technics and religion without structure and without qualities; by virtue of aesthetic activity, this neutral zone recovers a density and signification, while staying central and balanced; via aesthetic works, it recovers the reticular structure that stretched across the whole of the universe before the dissociation of magical thought.

While technical thought is made up of schemas, of figural elements without ground reality, and religious thought is made up of ground qualities and forces without figural structures, aesthetic thought combines figural structures and ground qualities. Instead of representing elementary functions, like technical thought, or the functions of totality, like religious thought, it holds elements and totality, as well as figure and ground by way of analogical relation; the aesthetic reticulation of the world is a network of analogies.

The aesthetic work is, in fact, linked not only to the world and to man, as a unique intermediate reality; it is also linked to other works, without conflating itself with them, without being in a material continuity with them, and keeping its own identity; the aesthetic universe is characterized by the possibility of passage from one work to another according to an essential analogical relation. Analogy is the foundation for the possibility of going from one term to another without a negation of the term by the succeeding one. It has been defined by Fr. de Solages[3] as an identity of relations [*rapports*], in order to distinguish it from resemblance, which in general would merely be a partial relation of identity. Complete analogy is, in fact, more than an identity of internal relations characterizing two realities; it is this identity of figural structures, but it is also an identity of the grounds of these two realities; on an even deeper level, it is even the identity of modes according to which, within the two beings, there is exchange and communication between the figural structure and the ground of reality; it is the identity of the coupling of figure and ground in two realities. Thus there is no true and complete analogy in the domain of purely technical thought, nor in that of purely religious thought; analogy applies to what one could call the fundamental operation of existence of beings, to that there is in them a coming-into-being that develops them by making figure and ground emerge; aesthetics grasps the manner in which beings appear and manifest themselves, i.e., by splitting into figure and ground; technical thought grasps only the figural structures of beings, which it identifies with its schemas; religious thought grasps only the ground schemas of the reality of these beings, according to which they are pure or impure, sacred or profane, saintly or sullied. This is why religious thought creates homogeneous categories and classes, like the pure and the impure, recognizing beings through inclusion within these classes or through exclusion from these classes; technical thought deconstructs and reconstructs the functioning of beings, elucidating their figural structures; technical thought operates, religious thought judges, aesthetic thought operates and judges at the same time, constructing structures and grasping the ground qualities of reality, in a related and complementary way, in the unity of each being: it recognizes unity at the level of the definite being, of the object of knowledge and the object of operation, rather than remaining, like technical thought, always below the level of unity, or, like religious thought, always above this level. 262

It is due to the fact that it respects the unity of definite beings that aesthetic thought has analogy as its fundamental structure; technical thought fragments and pluralizes beings because it privileges figural characteristics; religious thought incorporates them into a totality in which they are qualitatively and dynamically absorbed, thus becoming less than unity. In order to grasp beings at their level of unity, and to grasp them as multiple without annihilating the unity of each through division or incorporation, each being must be operated and judged as a

---

3. Bruno de Solages, a priest and member of the French Resistance who rejected Petain's collaboration and was sent to a concentration camp for advocating the church's stance against racism and for his activities in hiding Jews as rector of the Institut catholique de Toulouse. He is notably the author of *L'initiation métaphysique : l'univers, l'homme, Dieu, la connaissance* (1962); see also Bruno de Solage, *Dialogue sur l'analogie* (Paris: Aubier, 1946). [TN]

complete universe without excluding other universes: the constitutive relation of the being's coming-into-being, that which distinguishes and reunites figure and ground, must be able to transpose itself from one unity of being[4] to another unity of being. Aesthetic thought grasps beings as individuated and the world as a network of beings in a relation of analogy.

Thus, aesthetic thought isn't simply a remnant [*souvenir*] of magical thought; it is what maintains the unity of thought's coming-into-being as it splits into technics and religions, because it is what continues to grasp a being in its unity, whereas technical thought grasps the being below the level of its unity, and religious thought above it.

The aesthetic work is not a complete and absolute work; it is that which instructs as to how to move toward a complete work, which must be in the world and a part of the world as if it really belonged to the world, and not simply as a statue in a garden; while statues are beautiful in and of themselves, they do not make a house or a garden beautiful, a house and a garden are beautiful in their own right. It is by virtue of the garden that the statue can appear as beautiful, not the garden by virtue of the statue. It is with respect to the entire life of a man that an object can be beautiful. Furthermore, it is never the object strictly speaking that is beautiful: it is the encounter — which takes place about the object — between a real aspect of the world and a human gesture. Hence, it is possible for there to be no aesthetic object defined as aesthetic without nevertheless excluding the aesthetic feeling; the aesthetic object is in fact a mixture: it calls upon a certain human gesture, and furthermore contains an element of reality that, in order to satisfy this gesture and correspond to it, becomes the basis of this gesture, to which this gesture applies itself and in which it accomplishes itself. An aesthetic object that would be nothing more than objectively complementary relations would be nothing; lines would fail to be harmonious if they were pure relations; the separate objectivity of number and measure does not constitute beauty. A perfect circle is not beautiful insofar as it is a circle. But a certain curve can be beautiful even if it might be very difficult to find its mathematical formula. A line etching, representing a temple with very precise proportions merely relays an impression of boredom and stiffness; but the temple itself, worn by time and half crumbled, is more beautiful than the impeccable model of its erudite restoration. For the aesthetic object is not strictly speaking an object; it is also partially the depositary of a certain number of appeal aspects,[5]

4. *Unité d'être.* Here and throughout, *unité* in French means both "unity" and "unit." [TN]

5. "*Caractère d'appel*"; Simondon is referring here to a term coined initially by German social psychologist Kurt Lewin, a student of Carl Stumpf associated with the school of *Gestalt* theory and generally credited with inventing notions such as "group dynamics," "field theory," and "sensitivity training," as *Aufforderungscharakter* – literally, "invitation character." This

which are subject reality, gesture, waiting for the objective reality in which this gesture can exert and fulfill itself; the aesthetic object is both object and subject: it awaits the subject in order to put it into motion and solicit in it perception on the one hand and participation on the other. Participation consists of gestures, and perception gives these gestures a basis in objective reality. In the perfect model with its exact lines, indeed one finds all the objective elements represented [*figurés*], but there is no longer this echoing or appealing that gives objects the power to give rise to living gestures. Indeed, it is not the geometric proportions of the temple that give it this aspect of an appeal, but the fact that it exists in the world as a mass of stones, of coolness, darkness, and stability, which inflect our powers of effort or desire, our fear or our *élan* in a primary and preperceptive way. The qualitative load that is integrated in the world is what turns this block of stones into a stimulant [*moteur*] for our tendencies, rather than any geometric element that interests our perception. What remains on the sheet of paper where the reconstruction is drawn, is nothing more than geometric characters: they are cold and meaningless, because the arousal of tendencies has not been provoked before they are perceived. The work of art is aesthetic only to the extent that these geometric characteristics, these limits, receive and fix the qualitative flow. It is hardly useful to speak of magic to define this qualitative existence: it is biological as well as magical, it concerns the spontaneity of our tropisms, our primitive existence in the world before perception, as a being that does not yet grasp objects but directions, paths going upward or downward, toward darkness and toward light. In this sense the aesthetic object is ill-named, insofar as it evokes our tendencies; the object is an object only for

264

---

neologism, first introduced into English by J.F. Brown as "invitational character" (Brown, "The Methods of Kurt Lewin in the Psychology of Action and Affection," in *The Psychological Review* 36:3 (May 1929); 200-221), soon thereafter took its place in English and even eventually in German psychological vocabulary as "valence," now defined by the *Shorter O.E.D.* in psychology as "emotional force, significance, *spec.* the extent to which an individual is attracted or rejected by an object, event, or person," thanks to Donald Adams's translation of Lewin's "The Conflict between Aristotelian and Galileian Modes of Thought in Contemporary Psychology," in the *Journal of General Psychology* 5 (1931); 141-177. According to Lewin, *Auffordungscharakter* determines the meaning of an object with respect to the "direction" of the behavior of the subject to which it appeals in relation to the object. See Alfred J. Marrow, *The Practical Theorist, The Life and Work of Kurt Lewin* (New York: Basic Books, 1969), 55-63. Lewin's impact on French psychology seems to have come through the original German article as well as through Kurt Koffka's citation, which Simondon cites in his study on psychology as his source for the term which he translates as "*caractère d'appel*," and which he interprets in the direction of a sort of phenomenological realism: "Neither need, nor experience are sufficient to explain appeal aspects: there exists a 'silent' (implicit) organization of the object that belongs to it just like its color or its shape, and which makes the appeal character inherent in it [...] In this way, Koffka and Kurt Lewin abandon isomorphism: the system with a form is the whole [*l'ensemble*] constituted by the subject and his behavioral field, along with the entire network of forces connecting the sub-ensembles of the subject; there is no longer a physical field having its form, followed by a physiological field with own its form, and then finally psychological field having a third form, with these three independent forms being isomorphic among one another." The forces at play in appeal aspects "cannot be identified with those of classical mechanics," which "can be reductively explained in abstract vectorial terms; on the contrary, the tensions between Ego and objects have a nature and a specificity that are conserved in the effect; [...] the type of tension appeals to a certain response: attack, flight, approach, help, contempt, compassion." Simondon, *Sur la psychologie (1956-1967)* (Paris: Presses universitaires de France, 2015), 105-107. [TN] perception, when it is grasped as a localized *hic et nunc*. But it cannot be considered as an object in itself and prior to perception; aesthetic reality is pre-objective, in the sense that one can say the world is above all an object; the aesthetic object is an object at the end of a genesis that confers stability upon it and cuts it out; before this genesis there is a reality that is not yet objective, even though it is not subjective; it is a certain way of being in the world for the living thing, comprising the appeal, directions, tropisms in the strict sense of the word.

265   Real aesthetic feeling cannot be enslaved to an object; the construction of an aesthetic object is nothing more than a necessarily vain effort to recover the magic that has been forgotten; the true aesthetic function cannot be magical: it can only functionally be a memory and a re-enacting of magic; it is a magic going backward, a magic in reverse; whereas the initial magic is that through which the universe reticulates into singular points and singular moments, art is that through which a new reticulation emerges from out of science, morals, mysticism, and ritual and as a consequence of this new reticulation, there is the emergence of a real universe, in which the effort, which had been separated from itself, and which arose from the internal disjunction that technics and religion underwent, comes to completion, and as a consequence of these two expressions of magic, the initial effort of the structuration of the universe. Art reconstitutes the universe, or rather reconstitutes a universe, whereas magic starts from a universe in order to establish an already differentiated structure that carves the universe into domains charged with sense and power. Art aims at a universe on the basis of human effort and reconstitutes a unity. Art is thus the counterpart of magic, but it cannot entirely be this counterpart until after the two successive disjunctions.

    There are two partial forms of art: sacred art and profane art; art can intervene as a mediator between the mystical attitude and the ritual attitude; this art is like the act of a priest, without being what constitutes a priest; it rediscovers something of the mediator that has disappeared within the breaking apart that led to the appearance of the mystical attitude and the ritualistic attitude in place of religion. Sacred art is at once gesture and reality, object and subject, because it is at once the aesthetic attitude and the work; the work can exist only as something that is played; it comes from inspiration. Art is made of artistic activity and of the objectivized, actualized work; in this sense, there is mediation because there is celebration.

266   In the same way, profane art installs its object, namely the result of the artistic work, between theoretical knowledge and moral exigency; the beautiful is an intermediary between the true and the good, if one wishes to return to eclectic terminology. Like the tool, the aesthetic object is an intermediary between objective

structures and the subjective world; it is the mediator between knowledge and will. The aesthetic object concentrates and expresses aspects of both knowledge and will. Aesthetic expression and creation are at once knowledge and act. The aesthetic act comes to fulfillment, like knowledge, within itself; but aesthetic knowledge is mythical: it harbors a power of action; the aesthetic object is the result of an operation that is an intermediary between knowledge and action.

The aesthetic object, however, couldn't exist if aesthetic feeling didn't exist; it is merely what prepares, develops, and maintains natural aesthetic feeling, the sign of the fulfillment of a true encounter between the diverse elements of the world and the diverse gestures of the subject; every aesthetic object is either sacred or profane, while aesthetic feeling is at once sacred and profane: it pre-supposes the mediation of man at the same time as the mediation of the object; in aesthetic feeling, man is the priest of destiny just as the object is the object of destiny; destiny coincides with will.

This explains why there is in the work of art a stimulation of the tendencies and a presence of the sensible qualities, which are the tendencies' reference points. This also explains the definite structuration that gives the work of art the characteristics of the consistency of an object: the work of art calls upon both practical judgment and theoretical judgment.

But the aesthetic judgment is not necessarily the one made in the presence of a work of art; the work of art uses the prior natural existence of the spontaneous aesthetic judgment; furthermore, when the work of art presents a certain duration, aesthetic judgment is not given from the start in a state of fulfillment; there is a certain evolution of judgment that is initially more theoretical and ethical, and 267 which becomes more and more purely aesthetic as the completion of the work approaches; ancient tragedy offers a case of this evolution of modality during the course of the unfolding of the work: only the denouement corresponds to true aesthetic judgment; the duration that precedes it contains practical and theoretical judgments. Even in the aesthetic contemplation of a work that makes use of space rather than time, like painting or sculpture, there is a certain distinction of theoretical judgments and practical judgments, in a first moment of seeing, before the fusion and pure discovery of aesthetic feeling; one could even say that the work of art would always give off the experience of a certain disjunction between theoretical judgment and practical judgment if there wasn't the underlying solidity of technical judgment to sustain it: the work of art is a thing that has been made.

Aesthetic judgment generally remains a mixture of technical judgment and pure aesthetic judgment; of course there can be moments of pure aesthetic judgment throughout the unfolding of the perception of work of art; but one would think that aesthetic judgment would have the tendency to divide itself into theoretical judgment and practical judgment if the underlying work of art wasn't there to sustain the unity of apprehension, as a reality that has been made, and which in this respect has a real original unity. It is because of this presence of technical judgment in aesthetic apprehension that aesthetic judgment more easily appears in art than in life; in life, aesthetic judgment is extremely rare because an encounter is necessary that can only occur after a period of expectation and from an effort that polarizes the world, and if chance makes the occasional determinations of the world coincide with this universalized and concretized expectation; disappointment is infinitely more frequent than the aesthetic manifestation.

268 True aesthetic feeling, combining within itself the feelings of sacred art and of profane art, not only calls upon the aesthetic object (as profane art) or the human gesture of sacred art, but upon one and the other together: here man celebrates in the midst of a world of objects that have aesthetic value; the ancient tragic (aesthetic) is at once sacred and profane; it is what most closely approaches real life [*la vraie vie*] insofar as it arouses the feeling of the tragic, which is to say the feeling that seizes upon something within the human as a mediator; man's every gesture has a certain sacred aesthetic value; it intervenes between the totality of life and the world; it calls for participation. Destiny is this coincidence of the line of life and of the reality of the world via a network of gestures with exceptional value; every mediating gesture is aesthetic, even and perhaps essentially outside the work of art. The complete aesthetic gesture, at once sacred and profane, can hardly be found in the work of art, which is generally either sacred or profane. The complete aesthetic feeling is inseparable from the feeling of destiny; it does not take, from the sacred, the limitation to a definite domain of the real; it does not take, from the profane, the artificially objectivizing turn.

The sacred and the profane encounter each other in real life through the aesthetic impression; sacred art and profane art are only adjuvants of the complete and real aesthetic feeling, this feeling is not born from the work of art, be it sacred or profane, and it doesn't even require that the work of art be present at the moment in which the feeling manifests itself. The Romantics, who did not ask to be accompanied by the artificial work of art, found the true aesthetic feeling in life, without recourse to the explicit artwork that is made to be a work of art. Romanticism, however, is only one of the aspects of tragic thought, which attaches art to life, and, 269 for this reason, welds together the sacred and the profane. The mixture of genres in art is a direct consequence of Romanticism; but true Romantic aesthetic feeling is not in the artwork: it is in the attitudes of life. In classical art on the contrary, there is no reunion of sacred art and profane art: the forms of art are thus separated from one another, and true aesthetic feeling is in the work of art.

Established art [*art institué*] can achieve a partial meeting among thoughts that are closely related; but it cannot entirely ally religious thought with technical thought; by producing works of art, established art is nothing but a movement of a departure toward aesthetic existence, in which, for the subject, an encounter can take place, as the sign of a real accomplishment; true aesthetic feeling is of the domain of a reality experienced as reality; established art, artificial art is still nothing more than a preparation and a language for discovering true aesthetic feeling; true aesthetic feeling is as real and profound as magical thought; it comes from the real encounter between different particular modalities, recomposing the magical unity in itself, giving this unity back after a long disjunction. The aesthetic modality is thus a reunion of all modalities after their differentiation and separate development: through its power of unification, the aesthetic modality is what is the most functionally approximate to primitive magical thought. But the aesthetic impression can only be truly functionally equivalent to magic if it expresses a real encounter between different modal orders of thought, and is not simply the result of a factitious construction. The true sense and function of established art is to maintain the demand for unity throughout the modal orders that are differentiated in thought; if instituted art becomes aestheticism — in other words, if it gives and replaces a real and ultimate satisfaction considered as vitally experienced — it then becomes a filter that prevents true aesthetic feeling from appearing. 270

One can say that in this sense there is a continuous line going from magical thought to aesthetic thought, since in each of the modal orders of thought there are other underlying orders, which are the symbolic translation of the broken primitive unity. Thus within technics there is also a presence of its opposite, namely religious thought, which adds a certain sense of perfection to technics, namely technical beauty; in religious thought there is a desire to extend its mediation into the technical domain, and religious thought, while defending its norms against the penetration of the norms of technical thought, still tends toward a certain technicity, toward a definite regularity, toward forms that aestheticize it just as technical beauty aestheticizes technicity: within religious thought there is a religious beauty that represents the search for a complementary force whose aim is to recover the magical unity which has been broken, in the same way that there is within technical thought a search for a beauty through which the technical object becomes prestigious; the priest tends to be an artist just as the technical object tends to be a work of art: the two mediators aestheticize themselves in order to find an equilibrium that conforms to magical unity.

It must nevertheless be noted that, in the case of both religion and technicity, this premature aestheticizing tends toward a static satisfaction, toward a false completion prior to a complete specification; true technicity and true religion should not tend toward aestheticism, which maintains a rather facile magical unity through compensation, and thereby preserves magic and religion at a rather poorly developed level. The real development of thought requires the different attitudes of thought to be capable of detaching from each other and to even become antagonistic, for they cannot be simultaneously thought and developed by a single subject; they require, in fact, that a subject realize them and assume them in a profound, essential way, turning one of them into the principle of the subject's existence and 271 life. In order for an attitude to develop, thought must even be exchanged between several subjects and take on a temporal dimension, becoming a tradition and developing along a temporal line: hence the incorporation of a definite type into a social group as the foundation of its existence, as justification of its existence, as myth.

The more social or collective a thought becomes, however, the more it serves the individuals as a means of participation within the group, the more this thought also particularizes and weighs itself with historical elements and becomes stereotyped; a second function of aesthetic judgment thus becomes that of preparing for the communication between social groups representing the specialization of different types of thought. We have thus far presented the different modalities as if the human subject was individual and not collective; in reality, insofar as the subject is a collective being, art plays a preparatory role in realizing a commonality of the most diverse attitudes. There are technicians and priests, there are scholars and men of action: the original magical load that allows these men to have something in common and to find a way of exchanging their ideas resides in aesthetic intention. The category of the beautiful, in a specialized thought, is what announces that the needs of complementary thoughts are immanently and implicitly fulfilled by the very accomplishment of this specialized thought; the impression of beauty can hardly arise at the beginning of an effort, but only at its end, because this effort must first have followed its own direction, and it must furthermore arrive at the accomplishment of something it neither hoped to accomplish nor had already accomplished; beauty is gracious insofar as it is the accomplishment of what one didn't seek to accomplish, of that for which one didn't directly make an effort, and which was nevertheless obscurely felt as a complementary need, via a tendency toward totality. The tendency toward totality is the principle of the aesthetic search. But this same search begins a *progressus ad indefinitum,* because it is the will to perfection in each subject matter, whereas perfection aims precisely at domains 272 other than that in which it wishes to realize itself; under these conditions, the aesthetic quest cannot find stable norms, since it is driven by negative characteristics, which is to say by the feeling that a mode of thought discounts other modes of thought that are equally valid: the tendency of aesthetics is the endeavor to realize an equivalence of all other domains within a single determinate domain; the more particular and specialized a domain is, the more aesthetic requirements push for the construction of a perfect work, this perfection being a will to overcome itself in order to be equivalent to other domains and in order to realize them through an overabundance of this local accomplishment: as if this local perfection, radiating and overflowing superfluity, had the power to be what this domain is not.

Art is thus will to universality, will within a particular being to surpass its mode and realize all the modes within its own through a surpassing of its limits: perfection is not a fully realized normativity of limitation, but the discovery of an excellence so great, functioning within itself and reverberating within itself with such plenitude that it attains all other modes and would be capable of rendering them by way of an impoverishment. There is, in fact, an illusion in the aesthetic enterprise, since it is perhaps impossible that a determinate mode of thought can be equivalent to all the others by virtue of its perfection. Aesthetic intention, however, contains the affirmation of the possibility of this overcoming, of this equivalence or this mutual convertibility of excellences. Art is a quest for a concrete excellence, engaged in 273 each mode and with its sights set on finding the other modes through the movement of a mode within itself; this is how art is magical: it aims at finding modes without going outside a mode, only by dilating it, reworking it, and perfecting it. There is magic because there is a supposition of a reticular structure of the real universe; each mode magically exceeds itself while staying objectively within itself. This presupposes that the other modes are also subject to the same internal quest: it is not the stability of a mode that communicates with the stability of another, but excellence with excellence, aesthetic intention with aesthetic intention.

One could say, once again referring to the word transductivity, that art is what establishes the transductivity of the different modes in relation to each other; art is what remains non-modal in a mode, just as around an individual there remains a pre-individual reality associated with it and enabling it to communicate in the institution of the collective.

Aesthetic intention is what, to this extent, establishes a horizontal relation between different modes of thought. It is what allows the passing from one domain to another without recourse to a common genus; aesthetic intention harbors the transductive power that leads from one domain to another; it is the exigency of an overflowing and of the passage to the limit; it is the opposite of the sense of propriety, of the limit, of the essence contained in a definition, of the correlation between an extension and a comprehension.

Aesthetic intention in itself is already the exigency of totality, the quest for a whole reality. Without aesthetic intention, there would be an indefinite quest for the same realities within ever more narrow specializations; this is why aesthetic intention appears like a perpetual deviation on the basis of the central directions of a quest; in reality this deviation is a quest for the real continuity beneath the arbitrary fragmentation of domains.

274   Aesthetic intention enables the establishment of a transductive continuity attaching the modes to one another: one moves from the religious modes of thought to the modes of technical thought (perhaps it is better to say: from post-religious to post-technical thought), according to the following order: theological, mystical, practical, and theoretical; but this transductive relation is closed in on itself, so much so that it can only be grasped through spatial representation; one effectively goes from the theoretical to the theological as one moves from the mystical to the practical; there is a continuity between these two objective orders and between these two subjective orders. There is also continuity from a subjective order to an objective order within each of these two domains, the technical and the religious.

Aesthetic intention thus doesn't create, or at least should not create a specialized domain, which is that of art; but art, of course, develops into a domain and has an implicit internal finality: to preserve the transductive unity of a domain of reality that tends to separate itself by way of specialization. Art is a deep reaction against the loss of meaning and of the attachment to the whole [*ensemble*] of being in its destiny; it is not or must not be compensation, a reality occurring after the fact, but on the contrary a primitive unity, a preface to a development according to unity; art announces, prefigures, introduces, or completes, but it does not make real: it is the deep and unitary inspiration that begins and consecrates.

One could even ask oneself whether art, to the extent that it observes, is not also what somehow sums up [*résume*] and renders an ensemble of realities transposable to another temporal unit, to another moment in history. Art, in the celebration and final investiture that it brings about, transforms the fulfilled and localized reality *hic et nunc* into a reality that will be able to traverse time and space: it renders

human fulfillment non-finite; it is commonly said that art eternalizes different 275
realities; art, in fact, does not eternalize but renders transductive, giving a localized
and fulfilled reality the power to pass to other places and other moments. It does
not make eternal, but grants the power of rebirth and the capacity to fulfill oneself;
it leaves the seeds of quiddity; it gives the particular being realized *hic et nunc* the
power to have been itself but also to be itself one more time and a multitude of
other times; art loosens the bonds of ecceity [*eccéité*]; it multiplies ecceity, giving
identity the power to repeat itself without ceasing to be identity.

Art transgresses ontological limits, liberating itself with respect to being and
non-being: a being can become and repeat itself without negating itself and without refusing to have been, art is the power of iteration that doesn't negate the reality
of each new beginning; in this way it is magical. Because of it every reality, singular
in space and in time, is nevertheless a networked reality: this point is homologous
to an infinity of others that echo it and that are this point as well, however without
negating the ecceity [*eccéité*] of each node of the network: here, in this reticular
structure of the real, resides what one can call aesthetic mystery.

II. – *Technical thought, theoretical thought, practical thought*

The power of convergence of aesthetic activity fully exerts itself only at the level of
the relation between the primitive forms of technics and religions. But the power
of divergence contained within the autonomy of the development of technics and
religions creates a new order of modes of thinking that arise out of the splitting
of both technics and of religions, which are no longer at the natural level of aesthetic thought. With respect to these modes, aesthetic thought appears primitive;
it cannot make them converge through its own exercise, and its activity serves
only as a paradigm to orient and support the effort of philosophical thinking. Like 276
aesthetic thought, philosophical thinking is situated at the neutral point between
opposite phases; but its level is not that of the primary opposition resulting from
the phase shift of magical unity; it is that of the secondary opposition between the
results of the splitting in two of both technical thought and religious thought. It is
necessary to study this secondary splitting, and particularly that of technical activity, in order to know how philosophical thinking can efficiently and fully play its
post-aesthetic role of convergence, by applying itself to the becoming of technicity.

The level of the primary modalities of thought (technical, religious, and aesthetic) is characterized by the merely occasional use of communication and expression; aesthetic thought is of course capable of being communicated, and, technics and even religions can be learned, transmitted, and taught to a certain extent. However, it is through direct experience — which requires the subject to be put into a situation — that these primitive forms of thought are transmitted; the objects they create, their manifestations, can fall under the senses; but the schemas of thought, the impressions and norms that constitute these thoughts themselves and nourish them are not directly of the order of expression; one can learn a poem, contemplate a pictorial work, but this does not teach poetry or painting: what is essential in thought is not transmitted by expression, because these different types of thought are mediations between man and the world, and not encounters between subjects: they do not presuppose a modification of an intersubjective system.

277　　The secondary modalities of thought, on the contrary, presuppose communication and expression, they imply the possibility of a judgment, a node of expressive communication, and they contain, in the strict sense, modalities and attitudes of the subject as it finds itself before the content of its enunciation.

Now, technicity leads to certain types of judgments, and in particular theoretical judgment and practical judgment, or at least certain theoretical judgments and certain practical judgments.

It must in fact be noted that technicity is not alone in generating modalities of communicated thought by way of over-saturation and splitting; religious thought is also a basis for judgments.

The splitting of technical thought, like that of religious thought, arises from an oversaturated state of this thinking; at the primitive level, technical thought does not make judgments any more than religious thought does; judgments appear when the modalities differentiate themselves, for the modalities are modalities of thought, and in particular modalities of expression, before being modalities of judgment; judgment is simply the nodal point of expressive communication; it is as an instrument of communication that it has a modality, because the modality is defined by the type of expression; the modality is the expressive intention that envelops judgment, that precedes it and follows it. Modality is not contained in judgment; it makes it appear; judgment concretizes the modality of expression, but it does not exhaust it.

In technical activity, two opposing modalities emerge when an action fails, which is to say, when it forms an oversaturated, incompatible system with the world that it incorporates; if a single gesture always led to an identical result, if technical action were monovalent and without fissure, there wouldn't be any emergence of opposing modalities; technical thought would always be implicitly grasped by the efficacy of the accomplished act, and it would be indistinguishable from this act. But the failure of the technical gesture phase shifts the technical act into two opposing realities: one figural, made of schemas of action, habits, and structured gestures learned by man as a means, and the other a ground reality, 278 the qualities, dimensions, and powers of the world to which the technical gesture applies itself. This ground reality that undergirds the technical gesture is the dynamism of things, that through which they are productive, and that which gives them their fecundity, efficacy, and useable energy. Technics seeks the thing as power and not as structure, it seeks matter as a reservoir of tendencies, qualities, and its own particular virtues. It is nature as support and as auxiliary of action, as an adjuvant from which one expects efficacy, so that the gesture may prove effective. It is nature as a reserve of potentials, the φύσις [*physis*] that reveals its nature when it is lacking: it is something other than the schematic gesture of man; the gesture of man must be accomplished according to this productive nature in order to be technically efficient. This potentiality of nature, far richer than simple virtuality, is the foundation of the modality of possibility. Logical possibility is only the weakened reflection of the true virtuality of φύσις [*physis*], grasped and apprehended by way of its distinction from the human gesture, when the technical intention fails.

Virtuality, in turn, is a theoretical and objective modality, because it corresponds to that which is not within the power of man, and is nevertheless a power; it is pure power, absolute power.

Together with virtuality, the failure of technical action leads to the discovery of what subjectively corresponds to this virtuality, which is the possible as optative; the ensemble of schemas is an incomplete reality; the schemas of action are the beginnings of action, an incitement applied to the world so that an operation is carried out [*se réalise*]; this action is wanted, posited as desirable and already effectively desired insofar as man tends to carry it out; but it does not possess all its autonomy within itself, since human wish only has the status of a kernel of 279 action, and must encounter the virtuality of the world for there to be fulfillment: the practical optative corresponds to the theoretical virtual like a figural reality corresponds to a ground reality; the optative is the figure of the virtual. There is an implicit coupling here, directly given in the technical unity, prior to all modality. The emergence of these two modalities — one theoretical, the other practical — expresses the break in a first unity that was comprised of both knowledge and of action, concrete and complete technical thought.

But this is only one of the sources of practical thought and theoretical thought; the postulation of virtualities is not science, no more so than the possibility of schemas is practical thought; *physiology*[6] is a first attempt at science, but it is not science. It is essential to note that the notion of potential virtuality is always particular: it aims at a parceled elementary reality, taken piece by piece; it is not relative to the whole of the world; potential is only potential within a certain domain of the real and not within all of the real in the stable system that it forms: this character of virtuality, which often goes unnoticed, comes from technicity; technical action is effectively efficient or inefficient according to local powers; it must encounter in the *hic et nunc* a virtuality ready to actualize itself under the technical gesture; virtuality becomes integrated, localized, and particular. It is the objectively possible, just as the optative is the subjectively possible. It is thus natural that this modality of virtuality is what governs the inductive approach, aiming at the discovery of a truth through the accumulation of terms experienced one after the other. Induction is grounded, in its primitive forms, on virtuality and not on necessity; the truth obtained through induction could have been other than what it is; it is the adjunc-

280   tion of all these terms of virtuality that tend toward the real; one by one, they are virtual; but the system of all virtualities accumulated and linked to one another tends toward an equivalent of a basic stability, that of a virtual that is always available and present everywhere, corresponding to the "laws of nature." But before the laws of nature there are, to ground the primary inductive approach, the powers of nature, the φύσεις [*physeis*], the capacities to produce effects. Inductive thought is a thought that accumulates particular powers, rearranging them by similarities and domains, classifying the real according to the natural powers that can be discovered. In its first form, inductive thought prepares a general table of classification for technical action, destined to avoid technical failure by defining all the powers that the action can solicit, and by recognizing them deeply enough for it always to be possible to reach them, below the diversity of sensorial impressions.

Inductive thought is thus not defined only by its content; it is the form of theoretical thought that arises from out of the fragmentation [*l'éclatement*] of technics; it is, for method, the thought that goes from particular elements and experiences to the whole of the collection and to a general affirmation, seizing the validity of the general enunciation by way of the accumulation of the validity of particular experiences. For content, inductive thought is that which retains the qualities and the genetic powers of the world, such as the heavy and the light, the cold and the damp, the rigid and the flexible, the putrescible and the non-putrescible. All these

---

6.   This word is taken in the sense we give it when we speak of the "Ionian physiologists."

characteristics of things that the first inductive thought seeks are those that were implicated in technical operations: this does not in the least mean that theoretical inductive thought is a pragmatic thought, turned toward action and having as a sole aim to enable technical action; it is precisely the inverse: inductive thought comes from the failure of direct, parceled, localized technical action; this failure provokes the disjunction of the figural reality and the ground reality that was associated with it; inductive thought organizes the ground realities. But, even if it is not oriented toward action, inductive thought still bears the mark of its technical origin: in order for a ground reality, a φύσις [*physis*] to be grasped, it must have been associated with a definite technical operation: what induction retains, is what could have been evoked by the optative of action.

Through its failure, technical thought discovers that the world cannot be entirely incorporated into technics; if the world was made only of figural structures, a triumphant technics would never encounter any obstacles; but, beyond the figural structures that are homogeneous with the human gesture, there is another type of reality that negatively intervenes as an unconditional limit of the human gesture's efficacy. If water could rise to any height in the body of a pump, then the technics of the hydraulic engineer [*fontainier*] would suffice: the higher the height to be attained, the more perfect the construction of the body of the pump, the adjustment of the tubes, and the valve lapping should be: without a change of fields or the use of a new type of notions, there would only be proportionality between the importance of the result to be attained and the technical effort of construction. But when the water does not rise beyond a certain height in the suction pumps, then technical notions become inadequate; it is no longer the perfection of the technical object that is in question; the best hydraulic engineer cannot make the water rise above 10.33 meters; the world does not supply the technical gesture with a docile matter with no spontaneity; the world subjected to technical operation is not a neutral ground: it has counter-structures, opposing the figural technical schemas. Nevertheless, these obstructive powers of the world intervene within the axiomatic of each technics like an inexhaustible reserve of conditions that oversaturate this axiomatic as technics improve: a bucket wheel or an Archimedes screw do not encounter a counter-structure; but the elaborate art of the water works engineer capable of constructing a suction pump encounters this obstructive power. In particular it is worth noting that the new condition coming from this obstructive power is not homogeneous with the conditions of technical improvement: the conditions of technical improvement tend toward saturation through the concretization of the object systematizing itself as it perfects itself; but it is in addition to these conditions, and in a way that is not compatible with them, that the condition imposed by nature intervenes.

It is in order to recover the broken compatibility that technical thought splits into praxis and theory: theoretical thought which arises from technics is the thought at the heart of which it is possible to think the totality of the conditions of operation in a way that is once again homogeneous and coherent; through hydrostatics the system of conditions of the rising of water within the body of the pump can once again be homogeneous: since the rising of the water is explained by the difference in pressures exerted at the base and at the summit of the column, there is no longer a difference in kind between the previous technical conditions (leaking in the body of the pump allowing a residual pressure to subsist at the top of the water column, with a minimum pressure at the valve opening) and the previous non-technical conditions (the height of the column of liquid, the atmospheric pressure, and the liquid's vapor pressure); all the conditions are combined together in a homogeneous system of thought, centered around the notion of pressure, which is at once natural and technical; the technical failure leads thought to change levels, to ground a new axiomatic which, in a homogeneous way and by making them compatible, incorporates the figural schemas of technical operation and the representation of the limits imposed by nature upon the efficacy of these figural schemas in the technical gesture; it is the concept that is this new representation 283 establishing the notional compatibility. Science is conceptual not because it comes from technics, but because it is a system of compatibilities between the technical gestures and the limits imposed by the world upon these gestures; if it came directly from technics, then it would only consist of figural schemas, and not of concepts. The natural qualities, the φύσεις [physeis], understood as the material for technical gestures, constitute the most primitive type of concepts and mark the beginnings of inductive scientific thought.

The other result of this disjunction is the emergence of a practical thought that is not integrated within the real, but that is also made up of a collection of schemas, that are separated from each other from the very beginning. These *optatives,* freed from their application to the technical gesture, coordinate with each other like the objective virtualities of the world, and form a practical whole according to a process analogous to that of induction in theoretical knowledge. This is one of the bases of practical morals, with values such as those of efficacy of effort, of non-absurdity of action; such values must have been experienced and lived through an action integrated within the world, before being grouped and systematized; what is more, they cannot be completely systematized, because they lead to a plurality of

different values, just as inductive theoretical knowledge leads to a plurality of properties of things and laws of the real. Theoretical and practical thought, which come
from technics, remain pluralistic because of their inductive aspect. It is impossible
to say why there is one value for an action to be simple and easily accomplished,
and why there is another value for its efficiency; there is no analytical link between
ease and efficiency; and yet, there is a value for an action to be at once simple and
efficient. Only prior technical experience, really applied and integrated within the
hic et nunc, can provide the foundation of this pluralistic table of values of practical
morality. Constituted in practical thought, they are no longer technical norms, but   284
proceed from the experience of technical action meeting with failure, and rendering explicit both its objective foundations in inductive theoretical knowledge and
its subjective foundations in norms of practical morals.

The outcome of the split that occurs within religious thought, and which is the
correlate of the split that occurs within technics, opposes this pluralistic, inductive
and parceled aspect — which is pluralistic because it is empirical in its origin.
Religious thought also effectively splits into a theoretical mode and a practical
mode, when it becomes oversaturated to the point of incorporating too many subjective and objective elements to be able to remain compatible with itself as a
mediation between man and the world; it is essentially a collective subjective that
religious thought incorporates, translating the structures of society in its exigency
for universal representation. Charged with social inferences, religious thought can
no longer realize a mediation between man and the world; it thus splits into a representative exigency and a normative exigency, into a universal theological dogma
and a universal ethics. In these two specifications, it preserves what characterizes it
as religious thought, which is to say the demand for totality and for an unconditional unity given from the start.

Religious thought, like technical thought, effectively encounters limits to its
power, and these limits cannot be incorporated into its axiomatic. If religious
thought was applied to the world and to man without residue and without fissure, then the function of the respect of totality that it represents would never be
challenged; but other dimensions of totality emerge besides those coming from
the reticulation of primitive magic; the individual tendencies, and above all the
social groupings that develop and structure themselves over time have powers of
totality that cannot be mediated. Each city brings its own vision of the world, its
unconditional imperatives. Delphi cannot always remain a neutral ground when
cities develop into empires; there are powers in the universe that don't belong to   285
the ground aspects of the magical universe, and which are nevertheless also like

ground aspects. The oracle's power encounters another power that is of the same order, that ought to be compatible with it, and yet is not a part of primitive religious representation; it is a power that is not purely of ground; it has something structural, it particularizes the vision of the world; a city is a totality, an empire wants to be universal, and yet won't be completely; religious thought thus splits into theoretical and practical thought; while practical thought gives action a code, theoretical thought seeks to render the qualities and forces of the world compatible in a superior representation, grounding the θεωρία [theōría].

The theoretical knowledge that expresses religiosity seeks a systematic monistic representation of the universe and of man starting from the whole [le tout] in order to go toward the part, and from the whole [l'ensemble] of time in order to apprehend the instant in its particularity: it is monist and deductive, an essentially contemplative knowledge, whereas the theoretical knowledge coming from technics is operational; this knowledge is contemplative in the sense that the subject is in a situation of inferiority and posteriority with respect to the reality that is to be known; it doesn't constitute it through successive gestures as inductive knowledge does, by bringing order into an uncoordinated nature offering itself to his observation. For contemplative deductive knowledge, the effort of knowledge is only that of becoming aware of an already existing order, not that of an effective ordering; knowledge does not change being, and always remains partially insufficient for grasping being, which is prior to it and at the heart of which knowledge deploys itself like a reflection.

Indeed the use of number within the sciences appears to be of religious origin rather than of technical origin; indeed, number is basically structure that allows deduction and allows the grasping of a particular reality in its reference to the whole [l'ensemble], so as to integrate itself within it; it is the number of the philosophers, as defined by Plato who opposes the philosophical metretics to that of the merchants, a pure practical procedure that does not facilitate knowledge of the existence of the relations between beings, and between beings and the whole [le tout], conceived as cosmos. The ideal numbers are the structures that enable the relation of participation. Aristotle's critique of the number-ideas in the *Metaphysics* does not retain this eminently structural character of Plato's number-ideas, because Aristotle, following the schemas of inductive thinking, considers numbers through the operation of numbering; however, theoretical thought that makes use of numbers is essentially contemplative and of religious origin. It does not seek to count or measure beings, but to estimate what they are in their essence in relation to the totality of the world; this is why it seeks the essential structure of each particular thing within number. Religious thought, characterized by the function of totality and monist inspiration, is the second source of theoretical knowledge. It must be noted that its intention is to grasp universal figural realities, an order of the world, an economy of the whole [*le tout*] of being; in this quest, it is metaphysical and not physical, because it does not aim (as technical thought does by dissociating itself) at an inductive accumulation of local ground realities, powers or φύσεις [*physeis*]; it seeks the universal structural lines, the figure of the whole [*le tout*]. One can thus suppose that research coming from the deductive source of theoretical knowledge will never be able to meet the results of inductive research completely, since these approaches are grounded, one upon a ground reality and the other upon a figural reality.

Religious thought gives rise to an ethics of obligation at the level of the practical order, starting from a given unconditional principle and descending from this principle to the particular rules; there is an analogy between theoretical monism and the practical monism of forms of thought governed by religion; the order 287 of the world cannot be other than what it is; it is the opposite of virtuality; it is an actuality prior to all coming-into-being: the modality of deductive theoretical knowledge is necessity. The unconditional and unique aspect of the imperative in the practical realm, i.e., its categorical aspect, is what corresponds to the theoretical modality of necessity; this imperative orders. The way in which Kant presents the categorical imperative would be appropriate for defining the principle of ethics arising out of religion if Kant hadn't attached the categorical imperative to the universality of reason; the categorical religious imperative is categorical prior to being rational; it is everything at once, because the totality of being pre-exists all particular action and infinitely surpasses it, just as reality envelops the particular being who is the subject of moral action. The categorical character of the moral imperative translates the demand for totality, and the omnipotence of this demand in relation to the particularity of the being who acts; the categorical imperative is above all a respect for totality; it consists of the given and self-justifying character of ground reality. What the moral subject respects in the categorical imperative is the real as a totality that infinitely exceeds it, conditioning and justifying its action because it contains it; all particular action is taken from totality, and deploys itself on the ground of being and finds its normativity within it. It does not construct and does not modify it: it can only apply itself and conform to it. Here lies the second source of ethics, opposing itself to the technical source.

It can be said that there are two sources of theoretical thought and two sources of practical thought: technics and religion, taken at the moment in which they split because they are oversaturated and both have once again found a ground content and a figural content. Theoretical thought gathers both the ground content of technics and the figural content of religions: it thus becomes both inductive and deductive, operational and contemplative; practical thought gathers the figural content of technics and the ground content of religions, which provide it with both hypothetical norms and categorical norms, pluralism and monism.

A complete knowledge and a complete morality would be [found] at the point of convergence of the modes of thought coming from these two opposing sources in the theoretical order and in the practical order. However, what appears between these opposing exigencies is more of a conflict than a discovery of unity: neither theoretical thought nor practical thought succeeds in completely discovering a content that would truly be at the intersection between these two basic directions. Yet these directions act like normative powers by defining unique modalities that are capable of existing judgment by judgment, act by act.

In the theoretical order, this synthetic median modality is that of reality; the real is not what is primarily given; it is that in which the encounter between inductive knowledge and deductive knowledge would occur; it is the foundation of the possibility of this encounter, and the correlative foundation of the compatibility of a pluralist knowledge and a monist knowledge; the real is the synthesis of the virtual and the necessary, or rather the foundation of their compatibility; between inductive pluralism and deductive pluralism, it is the stability of the figure-ground relation taken as a complete reality.

Correlatively, in the practical order between the optative modality of practical thought coming from technics and the categorical imperative, there is the central moral category, situated at the intersection between the optative and obligation, between the pluralism of practical values and the monism of the categorical imperative; this modality has not received a name, because only the extreme terms have been noticed (hypothetical imperative and categorical imperative); yet in the practical order it corresponds to reality in the theoretical order; it aims at the optimum of action, by implicating a possible plurality of values and the unity of a norm of compatibility. The optimum is a character of action that renders the plurality of values and the unconditional demand for totality compatible. The optimum of action postulates a possible convergence of the hypothetical imperatives and the categorical imperative, and it constitutes this compatibility, just as the discovery of the structures of the real renders inductive pluralism compatible with deductive monism.

One could say that theoretical thought and practical thought constitute them-selves insofar as they realize a convergence toward the neutral center, finding once more an analog of primitive magical thought. However, theoretical unity and prac-tical unity, postulated by the existence of two median modalities of theoretical judgment and practical judgment (reality and optimum of action), allow for a hiatus to subsist between the theoretical order and the practical order; the primitive rupture dissociating the magical unity into figure and ground has been replaced by the bimodal character of thought, divided into theoretical and practical. Each mode, theory and practice, has figure and ground; but it is only together that they gather the complete heritage of primitive magical thought, the complete mode of man's being in the world. In order for the divergence within the coming-into-being of thought to be fully compensated, the distance between the theoretical order and the practical order would have to be overcome by a type of thinking that has a definitive capacity of synthesis, and is able to present itself as a functional analog of magic, and then of aesthetic activity; in other words, the work that aesthetic thought accomplishes at the level of the primitive opposition between technics and religion would have to be carried out anew at the level of the relation between theoretical and practical thought. This work is what philosophical reflection must fulfill [*accomplir*].

Now, in order for this philosophical work to be carried out [*s'accomplir*] the basis of this reflection must be firm and complete: in other words, the genesis of theoretical and practical forms of thought must be fully and completely ful-filled, so that the sense of the relation to be constituted can emerge. Philosophical thought, so as to be able to play its role of convergence, must first of all become aware of previous geneses, in order to grasp the modalities in their true signifi-cation, in order to be able to determine the true neutral center of philosophical thought; theoretical thought and practical thought are effectively always imper-fect and incomplete; it is their intention and their direction that one must grasp; this direction [*direction*], however, and this intention would not result from an examination of the actual content of each of these forms of thought; in order for philosophical effort to find the direction in which it must exert itself, what must be known is the direction [*sens*] of the coming-into-being of each form starting from its very origins. Philosophical thought has to return to an originary grasping of the coming-into-being at the end of which it (philosophical thought) intervenes as a force of convergence. It can even operate a conversion of technical thought and religious thought into relational modes before the dissociation that leads to the emergence of theoretical thought and practical thought; nothing effectively proves

290

that a viable synthesis can be established between these forms of thought if there isn't a common basic area pre-existing the dissociation, and reconnecting aesthetic thought to philosophy; this intermediate mode can be called culture; philosophy would thus be constructive and regulative of culture, translating the sense [*sens*] of religions and of technics into cultural content. In particular, its task would be to introduce new manifestations of technical thought and religious thought into culture: thus, culture would occupy the neutral point, accompanying the genesis of different forms of thought and preserving the result of the exercise of forces of convergence.

291   An effort of convergence can apply itself to recent forms of the elementary thinking of technics and of the thinking of totalities, the matrix of religions, because these two types of thought apply to the mediation, not only between the world and individual man, but between the geographical world and the human world; these two types of thought have human reality as their object, and elaborate themselves on the basis of this new weight; they refract human reality into different directions [*en des sens différents*]: this commonality of object can serve as a basis for the edification of a culture via the intermediary of philosophical reflection; there is a technics of man and every technics is to a certain extent a technics of man within a group, because man intervenes in the determination of the technical ensemble; the saturation of technical activity can lead to a structuration other than that of the fragmentation [*éclatement*] of thought into a theoretical mode and a practical mode; philosophical thought can allow for technical thought to remain more fully technical for a longer period of time, so as to attempt to relate the two opposing phases of man's being in the world before the dissociation of technical thought and religious thought; the task of philosophical thought would thus be to grasp afresh this coming-into-being, which means to slow it down in order to deepen its meaning and make it more fertile: the dissociation of fundamental phases of thought into theoretical modes and practical modes is perhaps premature; philosophical effort can preserve technicity and religiosity in order to discover their possible convergence at the endpoint of a genesis that wouldn't complete itself spontaneously without the genetic intention of philosophical effort. Philosophy would thus not only task itself with the discovery of genetic essences, but also their production.

CHAPTER III

TECHNICAL AND PHILOSOPHICAL THOUGHT

The opposition that exists between technics and religions, in an initial stage, is 293 inherent to the technics of the natural world's elaboration in the contrast they form with those religions that think the destiny of individual man. But there is a second stage of technics and of religion: after the elaboration of the natural world, technical thought turned itself toward the elaboration of the human world, which it analyzes and breaks down into elementary processes, and then reconstructs according to operational schemas, preserving the figural structures and setting aside the qualities and ground forces. To these technics of the human world correspond types of thought that also concern themselves with the human world, but taken in its totality. They aren't commonly called religions, because tradition reserves the name of religion for modes of thought that are contemporary with the technics of elaboration of the world; and yet, these modes of thought that assume the function of totality, in opposition with the technics applied to the human world, and which are the great political movements possessing a global reach, are indeed the functional analog of religions. But man's technics and social and political thought result from a new wave of splits [*dédoublements*] within magical thought. The old technics and religions were able to develop by taking advantage of the dissociation of the primitive magical universe considered almost exclusively as the natural 294 world; the human world remained enveloped by the primitive magical reticulation. Conversely, from the moment at which the man's technics broke away from this reticulation, and began considering man as technical matter, what emerged correlatively from this new break in the figure-ground relationship was a thinking that grasps human beings from below the level of unity (the technics of human management [*maniement*]) and another thinking grasping them from above the level of unity (social and political thoughts). As with the old technics and old religions that arose from out of the breaking of the natural world's magical reticulation, human technics and political thoughts proceed in opposite directions from one another; technics apply themselves to man by means of figural characteristics, pluralizing and studying him as a citizen, as a worker, and as a member of a familial community; it is indeed the figural elements that are retained by technics, and in particular criteria such as the integration into social groups, the cohesion of groups; they transform attitudes into structural elements like socio-metrics does when it transforms choices into the lines of the socio-gram. Rather than analyzing man, social and political thought classifies and judges him by putting him into categories defined by qualities and ground forces, just as religions classify and judge by placing each individual into the category of the sacred or the profane, or the pure or impure. And just as religions rebel against technics' profanation of the sacred aspect of certain places and moments, imposing respect for these places and moments on technics by way of prohibitions (for instance the observance of public holidays), in the same way, social and political thoughts, even when they oppose each other, limit the technics of man and oblige technics to respect his reality, as 295 if the technics of man were impious and disrespectful to the totality. The human world is thus represented in its elements by the technics of man and in its totality by the social and political concerns; but these two representations are not enough, because the human world can be grasped in its unity only at the neutral point; technics pluralize it, and political thought integrates it into a higher unity, that of the totality of humanity in its coming-into-being, where it loses its real unity in the same way that the individual loses its unity within a group.

However, the true level of human reality's individuation should be grasped by a thought that would be the analog for the human world of what aesthetic thought is for the natural world. This thought is not yet constituted, and it seems that it is philosophical thought that must constitute it. Aesthetic activity can be considered an implicit philosophy, but although aesthetic thought can apply itself to the human world, it seems unlikely that it would suffice to establish a stable and complete relation between man's technics and social and political thought. Indeed, this construction cannot be isolated because the human world is attached to the natural world. Man's technics emerged as a separate technics the moment when the technics of the elaboration of the natural world modified the social and political regimes through their rapid development. It is thus not only between man's technics and social and political thought that the relation must be established, but between all the elementary functions and all the functions of the whole [*les fonctions d'ensemble*], including the technics of man and the technics of the world, religious thought and social and political thought. Philosophical thought is appropriate for such an elaboration, because it can know the coming-into-being of the different forms of thought and establish a relation between the successive stages of genesis, in particular between the stage that carries out [*accomplit*] the break within the magical natural universe and the stage that carries out the dissociation within the magical human universe, and which is in the process of completing itself. Conversely, aes-   296 thetic thought is contemporary with each splitting into two even if it were possible to create a new aesthetics between man's technics and social and political thinking, a philosophical thinking, an aesthetics of aesthetics, would be required in order to attach these two successive aesthetics to each other. Philosophy would thus constitute the high neutral point of the coming-into-being of thought.

The philosophical effort thus finds itself faced with a unique task to be accomplished: the search for unity among the technical and non-technical modes of thought; but this task can take two different paths.

The first path would consist in preserving aesthetic activity as a model, and to attempt to bring about an aesthetics of the human world, so that the technics of the human world may encounter the functions of totality of this world, the concern with which is the animating force behind social and political thought. The second path would consist not in taking up technics nor those thoughts that assume the functions of totality in their original state, but rather only after their splitting into a theoretical mode and a practical mode, united in science and ethics. Now, the second path, which takes a longer detour, corresponds indeed to philosophical research, according to tradition as well as to the demands of a problematic; but it seems, within the current state of notions and methods, to lead to an impasse, to the point that Kant sought to distinguish the two domains of the theoretical and the practical, assigning an independent status to each of them. Already Descartes had sought to found [*fonder*] a provisional moral code, prior to the completion of theoretical knowledge. It can be asked whether the insoluble aspect of this problem of the relation between science and ethics does not rather come from the fact that science and ethics are not true, perfectly coherent and unified syntheses, but a rather unstable compromise between what technical thought contributes and   297 what religious thought contributes, which is to say between the demands of the knowledge of elements and of that of the functions of totality. In this case, one should rethink at its basis the genesis of the modes of thought, in the phase shift that opposes technics and religion, prior to the split that, within both technics and religion, leads to the emergence of the theoretical mode and the practical mode: philosophical thought, reflecting on technics and religion could perhaps discover a reflexive technology and an inspiration coming from religion which would directly and completely coincide with one another, rather than creating an intermediary space of incomplete and precarious relation, such as the one grounded in aesthetic activity.

This relation would be at once theoretical and practical, taken prior to the split into the theoretical mode and the practical mode. It would really and completely fulfill the role that aesthetic activity only partially fulfills, seeking to integrate technics and religion (here social and political thought is considered as being of the same order as religion, and capable of being treated like it) into a unique world that is both natural and human. For this integration to be possible, technical thought and religious thought would have to be at the level of unity, and no longer lower or higher than unity: the structures of plurality and of totality would have to be replaced by a network of analogically connected units.

The condition for this discovery is a deepening of the sense of technics and of the sense of religion that would lead to a reticular structuration of technics and religion. Technics and religion can coincide, not in the continuity of their content, but through a certain number of singular points belonging to both areas, and by establishing a third area through their coincidence, which is that of cultural reality.

298   Technical thought can be structured by the discovery of broader schemas than those of use in a determinate area. The pluralism of technics effectively results not only from the diversity of technical objects, but also from the human diversity of trades and areas of use. Technical objects with a wide variety of uses can have analogical schemas; the true elementary unity of the technical reality is not the practical object, but the concretized technical individual. It is possible to discover truly pure technical schemas (like those of the different modes of causality, conditioning, and command) through a reflection on these concretized technical individuals.

The reflexive effort applied to technics is characterized by the fact that a technics of all technics can develop through the generalization of schemas. Just as the pure sciences are defined, one can imagine founding a pure technics, or a general technology that would be very different from the theoretical sciences whose applications are translated into technics; it is indeed correct that a discovery in the area of the sciences can enable the birth of new technical devices [*dispositifs*]; but it is not directly, by deduction, that a scientific discovery becomes a technical device:

scientific discovery provides new conditions for technical research, but the effort of invention must nevertheless be exerted for the technical object to appear; in other words, scientific thought must become an operational schema or the basis for operational schemas. What one could call a pure technology resides, on the contrary, at the intersection of several sciences and traditional technical areas distributed among several occupations. Thus the schemas of circular action and their diverse regimes are not the property of any specific technics; they were first noticed and conceptually defined within the technics related to the transmission of information and automatism, because they play an important practical role within them, but even 299 before this, they had already played a role within technics such as thermal engine technology, and Maxwell had already studied them theoretically. And yet, every thought whose content covers a plurality of technics, or at the very least applies to an open plurality of technics, exceeds for that very reason the technical domain. Certain processes included within the functioning of the nervous system can be thought of through schemas of recurrent causality as well as certain natural phenomena; the schema of relaxation for instance is always identical to itself, whether it is applied to a technical apparatus, to the functioning of an intermittent fountain or to the phenomenon of trembling in Parkinson's disease. A general theory of causalities and of conditionings exceeds the specificity of a domain, even if the conceptual origins of this theory come from a particular technics. The schemas of a generalized technology therefore rise above the distinct technical object; in particular, they allow for adequately thinking the relation between technical objects and the natural world, which is to say they allow for the assurance of the integration of technics into the world in a way that goes beyond empiricism. The technical object, placed into the middle of a body [*faisceau*] of actions and reactions whose interplay is predicted and can be calculated, is no longer that object separated from the world, resulting from a break within the primitive structuration of the magical world; the figure-ground relation, broken by technical objectivation, is once again found within general technology; because of this, the technical object is invented according to the milieu into which it must be integrated, and the particular technical schema reflects and integrates the characteristics of the natural world; technical thought extends itself by incorporating the demands and the mode of being of the milieu associated with the technical individual.

Thus, to the extent that a polytechnic technology replaces the separate technics, the technical realities themselves, in their realized objectivity, take on a network 300 structure; they are related to one another, rather than being self-sufficient like works by artisans, and they are in relation to the world which they bind into the mesh of their key-points: tools are free and abstract, always transportable everywhere one goes, but technical ensembles are true networks concretely attached to the natural world; a dam cannot be built just anywhere, nor can a solar furnace. Some notions of traditional culture appear to suppose that the development of technics causes the disappearance of the particular aspect of each place and region, leading to the loss of customs and local artisanal inflections; in reality, technical development creates a far more important and much more firmly rooted concretization than the one it destroys; an artisanal custom, like a regional costume, can, by simple influence, be transported from one place to another; it is only rooted in the human world; conversely, a technical ensemble is profoundly rooted in the natural milieu. There are no coal mines in primary terrains.

So it is that some high points of the natural, technical, and human world are constituted; it is the ensemble, the interconnection of these high points that makes this polytechnic — both natural and human — universe; the structures of this reticulation become social and political. In existence, technics are separate neither for the natural world nor for the human world. And yet, for technical thought, they remain as if they were separate due to the fact that no thought has been sufficiently developed in order to allow for the theorization of this reticulation of concrete technical ensembles. This theorization is the task that befalls philosophical thought, for there is a new reality here that is not yet represented in culture. Beyond the technical determinations and norms, one would have to discover polytechnic 301 and technological determinations and norms. A world of the plurality of technics exists that has its own structures, and which ought to find representations within the content of culture would be adequate to it; the general term network, commonly employed to designate the interconnecting structures of electrical energy, telephones, railways, and roads, is far too imprecise and does not account for the particular regimes of causality and conditioning that exist in these networks, and that functionally attach them to the human world and to the natural world, as a concrete mediation between these two worlds.

The introduction of adequate representations of technical objects into culture would result in the key-points of technical networks becoming real terms of reference for the ensemble of human groups, whereas they currently are only key terms for those who understand them, which is to say for the technicians of each specialty; for other men, they only have a practical value, and correspond to very confused concepts; technical ensembles introduce themselves into the world as if they had no natural or human right of belonging, while a mountain or promontory, which have less concrete regulatory power than some technical ensembles, are known by all men of a region and belong to the representation of the world.

And yet, one can wonder to what extent the creation of a general technology brings technics closer to religion; the recognition of the genuine [*véritable*] complex operational schemas and of the integration of technical ensembles would not be enough to enable this rapprochement if there weren't, along with a theoretical awareness of processes also a normative value contained in them. Indeed, the reticular structures of integrated technics are no longer mere means available for an action and abstractly transportable anywhere, utilizable at any moment; one   302 changes tools and instruments, one can construct or repair a tool oneself, but one cannot change the network, one doesn't construct a network of one's own: one can only connect to a network, adapt to it, participate in it; the network dominates and frames [*enserre*] the action of the individual, it even dominates each technical ensemble. Whence a form of participation in the natural world and in the human world that gives an incorruptible collective normativity to technical activity; it is no longer only a slightly abstract solidarity of trades as evoked by Sully Prudhomme (the solidarity of specialists, the bricklayer, the baker), but an extremely concrete and actual solidarity, existing instant by instant throughout the interplay of multiple conditionings; through the technical networks, the human world acquires a high degree of internal resonance. The powers, forces, and potentials that drive toward action exist in the reticular technical world in the same way in which they might have existed in the primitive magical universe: technicity is a part of the world, it is not only an ensemble of means, but an ensemble of conditionings of action and of incitements to act; the tool or instrument doesn't have normative power because it is permanently available to the individual; technical networks take on more normative power as the internal resonance of human activity throughout technical realities becomes greater.

However, the valorization of technical ensembles and their normative value entail a very particular form of respect, which has in view pure technicity in itself. It is this form of respect, founded on the knowledge of technical reality, and not on the prestige of the imagination, which can penetrate culture. A large highway, at the edge of a big city, imposes this form of respect; moreover, a harbor, the rail traffic signal regulation center, or the control tower of an aerodrome impose this same form of respect: the key-points of a network possess this power, insofar as they are key-points, and not because of the direct prestige of the technical objects   303 they contain. So it was that the clock of the Paris Observatory, about ten years ago, was slightly disrupted by the tumultuous visit of science students passing by it on their way to the catacombs; the uproar caused by this violation of the technical sacred was, at that moment, rather considerable. Now if that same clock had been placed in a teaching laboratory, and it had been intentionally set wrong in order to show the interplay of the self-regulation of its functioning, then no emotion corresponding to the violation of the sacred would have been felt; it is in fact because the Observatory's clock is the key point of a network (it emits the time signals by radio) that its disturbance is scandalous; it is not because of the practical danger that this disturbance would have represented, because it is too small to be severe enough to lead to important errors by ships on sea. Here we bear witness to a profanation properly speaking, independent of the practical consequences it could entail; it is the stability of a system of references that is compromised. It is likely, by the way, that to have attempted such a profanation would not have occurred to humanities students, since for them the Observatory's clock doesn't have the same normative value; it is not sacred, because it is not known according to its technical essence, and is not represented by adequate concepts in their culture. These forms of respect and of disrespect manifest, within the technicity that is integrated into the natural and human world, the inherence of values surpassing utility; thought that recognizes the nature of technical reality is that which, going beyond separate objects — utensils — according to Heidegger's expression, discovers the essence and reach of technical organization, beyond separate objects and specialized occupations.

304   Traditional religious thought seems to find a means of becoming aware of itself in an attitude of struggle against new technics. In fact, it is not technics themselves that are targeted, but the type of civilization that is contemporary to these technics and that abandons not only their traditional religions, but also the old technics that were their contemporaries. This opposition is distorted at its very foundation because current technics should be coupled to social and political thoughts, and not to religions, which are not their contemporaries. It is only after the realization of the coupling of technics and religions of the same era that the continuity of successive stages can be perceived, but not by way of the opposition of the phase of one era with that of another.

If one considers the social and political thoughts of our time, which are contemporary to the recent development of technics, it can be seen that they bring religion's aspect of absolute universality down to a dimension that conforms to an integration into the natural and human world; every social and political doctrine no doubt tends to present itself as an absolute, as unconditionally valid, beyond the *hic et nunc*; social and political thought, nevertheless, accepts positing current and concrete problems; in the same way as technical thought being developed does, social and political thought leads to a reticular representation of the world, with key-points and essential moments; it applies to technical reality by treating it as more than a simple means, and indeed grasps technical reality at the level of the reticulation of integration into the natural and human world. Three important recent social and political doctrines have thus incorporated, each in an original way, a representation and valorization of integrated technics; National Socialist thought is attached to a certain conception linking the destiny of a people to a technical expansion, even thinking the role of neighboring peoples as a function of this master expansion; the American democratic doctrine has a certain definition of technical progress and of its incorporation into civilization; the notion of the standard of living, which is a social one, and constitutes a cultural reality, has a con-   305 tent wherein important terms are technological (not only the possession of this or that instrument or utensil, but the fact of knowing how to use this or that network, of knowing how to be functionally connected). The Communist Marxist doctrine finally, in its lived and realized aspects, considers technical development to be an essential aspect of the social and political effort to be made; it gains self-awareness through the use of tractors, the foundation of factories. At the political level, the self-awareness the great nations have of themselves contains a representation not only of their technical level (which would only be an estimation of power), but of their integration through the intermediary of technical reality into the entire current universe. A change in technics entails a modification of what one could call the political constellation of the universe: the key-points move on the surface of the world; coal is less important today than at the dawn of the Great War; but oil is more important. These structures are more stable than the economic structures that govern them: despite a great number of economical modifications, certain passageways toward mineral resources have remained stable since the times of Roman conquest. Social and political thought becomes integrated into the world according to a certain number of outstanding points, problematic points that coincide with the points of integration of technicity envisaged as a network.

By this we do not mean to say that social and political structures limit themselves to expressing the state of the economy which is itself determined by the state of technicity, what we mean to say is that the distribution and integration of key-points of social and political thought in the world at least partially coincides with the distribution and integration of the technical key-points, and that this coinciding becomes all the more perfect as technics becomes increasingly integrated within the universe, in the form of fixed ensembles, attached to one another,   306 constraining [*enserrant*] human individuals into the links they determine.

However, such a formal rapprochement of the structures of political thought and of technical thought cannot resolve the problem of the relationship of technics with forms of non-technical thought. It is effectively at the price of a certain renouncement of universality that social and political thought manages to make its structures coincide with those of technical thought, and particularly of technical thought applied to the human world; social and political thought manages to coincide well enough with the representations of commerce, with import and export, in other words with economic realities that are the result of the existence of technics, but which translate the manner in which technics are used by human groups; these modes of utilization of technics by human groups are themselves subject to technics that no longer apply to the natural world, but rather to the human world, and that don't produce technical objects or technical ensembles, unless one can consider the means of advertising or the retail organizations as such. Currently, it could therefore be said that the agreement between technical thought and non-technical thought is only possible at the price of very great simplification and abstraction, as much in the technical domain as in the non-technical domain.

On the one hand, this simplification essentially consists in establishing a break between the technics of the natural world and the technics of the human world, and on the other hand, in a break between religious thought and social and political thought. Through this rupture, by virtue of the abandonment of the technical demands of the natural world, the technics of the human world, rather than being constrained to remain within an elementary plurality, below a true unity, could think that they grasp a true unity in the generality of groups, of the crowd, and of public opinion; in reality, they continue to apply elementary thought to general realities, for instance, by studying the *mass media*[7] as if they were distinct from the concrete reality of the groups in which they operate; the rupture between figure and ground subsists in the technics of the human world, and it is even particularly obvious, but it goes unperceived in the implementation of technics, because these technics, in order to act, seek precisely what one could call the ground figures, namely those that are the least formalized and the least institutionalized. Despite this aspect, they nevertheless remain figural realities, and not a whole and complete reality.

This same insufficiency manifests itself within socio-political thought which remains the intermediary between the true consideration of totalities, characterizing real religious thought, non-politicized or socialized by the influence of a group, and the mythological application to the expression of the necessities of a moment

---

7. English in original. [TN]

or a group: it is generally the mythology of a group that is erected as a universaliz-able doctrine; because of its claim to universality, of that which is only universal in regard to its origins and intent, social and political thinking is a combat thinking. Henceforth, it becomes easy to understand why the distance between the technics of human management and social and political thought is not so great: a political movement can use the technics of advertising converted into means for propaganda, just as a definite technics of human management leads to a social and political option. But this encounter, this mutual complicity, can only exist at the price of abandoning fidelity to the elementary functions, characterizing true $\quad$308 technicity, and the correlative abandonment of the mission to represent the func-tions of totality characteristic of religious thought. The alliance of an ensemble of procedures and of a mythology is not the encounter of technicity and of respect with regard to totality.

This is why philosophical thought must maintain the continuity that exists between the successive stages of technical thought and religious thought, all the way to social and political thinking. Technicity must be maintained starting with the technics applied to the natural world all the way up to including those address-ing the human world, just as the concern with totality must be maintained starting from religions and progressing all the way up to including social and political thought. Without this continuity, without this real unity of the coming-into-being of technics and of the thoughts related to the function of totality, a false dialogue is established between the forms relating to the natural world and those relating to the human world; for instance, the technics of human management are only an additional variable within industrial technics (*scientific management*),[8] or else tradi-tional religious thought chooses a current social and political thinking that adopts the worldview closest to it, and thereby deprives itself of its power of universality.

Because of its object, this study does not propose to deal with the problem of establishing continuity between the religious and social and political forms of thought; it must nevertheless deal with this problem insofar as this effort is sym-metrical with the effort by which the technics of the world must be brought closer to those of man.

And yet, if the technics of man fail in their function of the analysis of elements, and act generally through empirical procedures (which is what statistical concep-tualism expresses, developing within a comfortable nominalism), this is because they accept detaching themselves from the real object, the element, individual, or ensemble. There can be no true technics separate from the human world; the

---

8.   English in original. [TN]

309 technics of the human world must have an objective material, they cannot be purely psychological, lest they become procedures, in other words, it is by way of an enlargement of technical ensembles comprising an integration into both the natural world and the human world that one can subsequently act upon the human world, that is, through, and according to, this natural and human ensemble: as mediation between the natural world and the human world, technical thought can act upon the human world only via the intermediary of this mediation. Human reality can be the object of technics only when it is already engaged within a technical relation. There is no legitimate technics other than that of technical reality; technical thought must develop the network of relational points between man and the world, by becoming a technology, which is to say, a second degree technics that deals with organizing these relational points. But there cannot be any legitimate application of technical thought into a non-technical reality, for instance into what one could call the natural and spontaneous human world: technology can only develop within an already technical reality. Reflexive thought must promote technology, but it mustn't attempt to apply technical schemas and procedures outside the domain of technical reality.

To put it another way, it is not human reality (and in particular that which in human reality can be modified, namely culture: the active intermediary between successive generations, the concurrent [*simultanés*] human groups and successive or concurrent individuals) that must be incorporated into technics as a workable material; it is culture, considered as a lived totality, that must incorporate the technical ensembles by knowing their nature, in order to be able to regulate human life according to these technical ensembles. Culture must remain above all technics, but it must incorporate into its content the knowledge and intuition of genuine

310 technical schemas. Culture is that through which man regulates his relation with the world and with himself; and yet if culture were not to incorporate technology, it would contain an opaque zone and wouldn't be able to contribute its regulative normativity to the coupling of man and the world. For in this coupling of man and the world, which is that of technical ensembles, there are schemas of activity and conditioning that can be clearly thought only by virtue of concepts defined by a reflexive but direct study. Culture must be contemporary with technics. Culture must reshape itself [*se reformer*] and must once again take up its content stage by stage. If culture is only traditional, then it is false, because it implicitly and spontaneously contains a regulative representation of technics of a certain era; and it falsely brings this regulative representation into a world to which it cannot apply. Therefore, the confusion of technical realities with utensils is a cultural stereotype,

founded on the normative notion of utility that is at once valorizing and devaluing. But this notion of utensil and of utility is inadequate to the effective and actual role of technical ensembles within the human world; it thus cannot be regulative in an effective way.

Deprived of the contribution of cultural regulation, which passes through the intermediary of an adequate representation of technical realities, this coupling of man and the world develops in isolation, in a non-integrated, anomical manner. As a counterpoint, this development of technical realities enveloping man without regulation justifies, at least apparently, culture's implicit mistrust of technics; a self-justifying culture develops in human milieus promoting a technics, while general culture becomes inhibiting, but not regulative, of all technics.

Now, the philosophical and notional awareness of technical reality is necessary for the creation of a cultural content incorporating technics, but it is not enough. 311 Nothing effectively proves that technical reality can be adequately known through concepts; conceptual knowledge can indeed designate and cover technical reality at the level of separate technical objects, which can be classified according to structures and usages. But only with great difficulty can it lead into knowledge of technical ensembles. To acquire this knowledge, the human being must really be put into a situation, since it is a mode of existence that he must experience. The tool, instrument, or isolated machine can be *perceived* by a subject who remains detached from them. But the technical ensemble cannot be grasped by intuition, for it cannot be considered a detached, abstract, or manipulable object at man's disposal. It corresponds to an experience of existence [*épreuve*] and a situation, it is tied by reciprocal action with the subject.

In the same way one used to consider journeys as a means for acquiring culture, because they constituted a mode of placing man into a situation, one should also consider the technical experiences of being placed into a situation with respect to an ensemble, with effective responsibility, as having cultural value. To put it another way, every human being should to a certain extent take part in technical ensembles, that is, take on a responsibility, a definite task with respect to such an ensemble and be connected with a network of universal technics. Furthermore, individual man should not simply experience a single kind of technical ensemble, but rather a plurality of them, just as a traveler will have to encounter several peoples, and experience their mores.

However, this kind of experience must be conceived more as a way of experiencing the situating of each type of technics and ensemble of technics, than as an effort to participate in the condition of man in each of the technics: for in each 312 technics there are technicians, unskilled laborers, workers, managers, and to the extent that conditions are strictly social, they can be rather analogous, at each level, in the different technics. It is the particular situating in the technical network that must be experienced, insofar as it places man in the presence of and within a series of actions and processes that he does not direct alone, but in which he participates.

The philosopher, comparable in this role to the artist, can help in raising awareness of the situation within the technical ensemble, by reflecting it within himself and by expressing it; but, again just as the artist, all he can do is be the one who solicits an intuition in others, once a definite sensitivity has been awakened and allows the grasping of the sense of a real experience.

However, we must note that art, as a means of expression and as awakening of the cultural awareness of technical ensembles, is limited; art goes through αἴσθησις [aísthēsis] and is therefore naturally inclined to grasp the object, tool, instrument, or machine; but true technicity, that which can be integrated into culture, is not in the manifest. All the prestigious color photographs of sparks, of fumes, all the recordings of noise, sounds, or images, generally remain a use [*exploitation*] of technical reality and not a revelation of this reality. Technical reality must be thought, and even be known through participation in its schemas of action; aesthetic feeling can emerge, but only after this intervention of real intuition and participation and not as a fruit of a mere spectacle: every technical spectacle remains puerile and incomplete if it is not preceded by the integration into the technical ensemble.

The intuitions of technical participation, however, are not opposed to the forces and qualities of religious and social and political thought. Social and political thought is continuous with respect to religious thought when it is not an actual and already realized totality properly speaking (for totality is what it is, it is an absolute and cannot push toward action), but instead is the latency of broader ensembles undergirding actual structures, and the validity of this announcement of new structures; social and political thought expresses the relation of totality with respect to the part, of virtual totality with respect to the actual part. It expresses the function of relative totality, whereas religions express the function of absolute totality, and it expresses the function of virtual totality, whereas religions express the function of actual totality. And yet, there can be a complementary relation between the intuitions of integration into technical ensembles and social and political intuitions, because technical intuitions express the result of history and of the conditioning of life, of the *hic et nunc*, whereas the social and political intuitions are a project going toward the future, the active expression of potentials. Social and political thoughts are the expression of tendencies and forces that exceed all actual given structure; the intuitions relating to technical ensembles express what humanity has done, what is done, and what is structured because done, accomplished. Figural power can thus remain invested in technics and ground power in social and political thought, insofar as figural reality is what is given in the system of actuality whereas ground power contains potentials and keeps the becoming in reserve. While impossible at the level of the relation between elementary technical objectivity and universal religious thought, the relation once more becomes possible when it establishes itself between technical ensembles, which is the expression of actuality, and social and political thought, which is the expression of virtuality. There is a compatibility between actuality and virtuality through the real coming-into-being, whose meaning [*sens*] is stretched taught between this actuality and this virtuality. Philosophical thought grasps the correlation between actuality and   314 virtuality, and it maintains it by establishing the coherence of this relation.

It is thus the sense [*sens*] of coming-into-being, the ability of technics to engender the coming-into-being of both the natural and the human world, that makes elementary intuition and the intuition of the ensemble compatible; technical intuition, at the level of ensembles, expresses coming-into-being as both basis and result obtained; social and political intuition is the integration of tendencies, the expression of virtualites and forces of coming-into-being, in the same reality. At the level of technical thought attached to tools, and of universalizing religious thought, there cannot be any direct encounter between the two types of thought, because the mediation of coming-into-being is not possible; each tool, each separate technics that is capable of manipulating tools present themselves as stable and definitive. Universalizing religious thought also presents itself as stable and definitive, with reference to an atemporal ground. Conversely, the introduction of technicity to ensembles which situate man as organizer or as element makes technics evolve; to the same extent and at the same time, the evolving aspect of human groups becomes conscious and this consciousness creates socio-political thought. Both born from coming-into-being, one expressing the definite past serving as its basis and the other the possible future serving as its goal, the technical thought of ensembles and social and political thought are coupled through their conditions of origin and their points of integration into the world.

Thus it is within the perspective of permanent change within technical and socio-political structures that technical thought and socio-political thought can coincide. Elementary technicity, the one that animates the thought of artisans, and religiosity with a universal basis, the one that is contemporary with the first development of technics, can serve as paradigm for thinking the coming-into-being of   315 technical ensembles and for thinking the coming-into-being of totalities; without the norm of elementary technics and that of universal religiosity, the technical thought of ensembles in a process of coming-into-being and the social and political thought of evolving communities would lose their reciprocal tension; the thought of technical ensembles needs to be inspired by that of elements, and that of the coming-into-being of the human world by the function of totality, in order for these two forms of thought, which must meet analogically but which must not be confused with each other, to preserve their autonomy and not enslave one another. Because the functional totality of thought coming from the primitive relation with the world must be maintained by the real bipolarity of the primitive phase shift's results; culture is directed by this bipolarity; it develops between technical thought and religious thought; it is culture that links the lived understanding [*compréhension*] of the technicity of ensembles with that of the human groups represented in socio-political thought.

The past — i.e., the first forms of technical thought and of religious thought, at the level of the first splitting in two of magical thought, as well as aesthetic activity placed at the neutral point of this first split — must be preserved as cultural content, i.e., as a foundation for providing the norms for current thought, but it is only as cultural content that they must be preserved; it would be a transgression [*faute*] against coming-into-being to want to substitute the representation of elements, tools or instruments for that of the current technicity of ensembles; for technicity, in its current lived reality, no longer simply resides at the level of elements, but also and essentially at the level of ensembles; today ensembles are depositaries of technicity in the same way that the fragmentation into elements had been prior to this; thought must start from the knowledge of the technicity of elements, re-situated in the past, in order to grasp the technicity of ensembles in their reality, for it effectively results from it: thought must go from the cultural to the actual in order to understand the actual in its reality. Furthermore, religious thought is a permanent reminder of the sense of totality, and culture must renew the rootedness of socio-political thought in universalized religious thought, proceeding from the cultural to the virtual, in order to grasp and promote the virtual in its value.

316

Now, the non-cultural in technics is the uniqueness of each determinate technics, tending to impose its norms, schemas, and particular vocabulary; technics, in order to be grasped in their real essence which alone is cultural, must be presented and experienced as a cluster [*faisceau*] of plurality; this plurality is a part of the technical condition, which grasps the elements. Religious thought, inversely, must be seized as unconditional unity, in itself; what is contrary to culture, in religions, is their possible plurality, which is to say the confrontation of determinate religious traditions; and yet, since religions, as traditions, are necessarily rooted, culture must create a superstructure on the basis of which the different religions appear in their unity as religions; it is the meaning [*sens*] of ecumenism, which is the condition of the integration of religions into culture, the condition of religions' fecundity in the direction [*sens*] of culture; it is perhaps uncertain whether there really can be open religions, or whether the opposition between closed religions and open religions is as clear as the one Bergson establishes; but the openness of religions is a function that is common to different religions, each to a certain extent closed in on itself.

It was hardly possible for ecumenism to be constructed in a distant past, for it can constitute itself only by means of a reflexive thinking wanting to ground culture; it is essentially, and in itself, philosophical work; it necessitates becoming aware of the deep sense of religions, which can occur only by re-situating them within the coming-into-being of thought on the basis of primitive magic. To this day, limited ecumenisms (as within Christianity) have arisen, but it is a universal ecumenism that philosophical reflection must develop so that religious reality integrates itself into culture. 317

The institution of a technology has the same signification as that of ecumenism, but its consequence is making one grasp the true elementary particularity of technical objects, on the basis of a general normalization of the common vocabulary and notions, replacing the false specificity of trade terms, caused by use and not by the essence proper to the elements; technology is that on the basis of which the plurality of technical objects, which is the depositary of primitive technics, serves as the basis for the constitution of technical ensembles. Ecumenism is that on the basis of which the universalizing unicity of religious thought, which is the depositary of the function of primitive totality, serves as a basis for socio-political thought. Technology accomplishes, on the basis of plurality, a conversion toward unity, whereas ecumenism, first of all grasping unity, accomplishes or allows for the accomplishment of a possible conversion toward a plurality of social and political integration. The conscious grasping of this function of plurality and of the function of unity are necessary as a basis, so that mediation, at the level of this encounter between the status of plurality and the status of superiority with respect to unity that is realized by the structure of reticulation, can be possible at the neutral point of the coming-into-being of thought.

However, in order for philosophical thought to be able to perform the integration of the sense of technics into culture, it is not enough that it applies itself to culture outside philosophy strictly speaking, as it could accomplish a limited task

out of duty; because of the reflexivity of thought any philosophical activity is also a reformation of the mode of knowledge, and reverberates within the theory of knowledge. In turn, becoming aware of the genetic aspect of technicity must lead
318 philosophical thought to address the problem of the relations between concept, intuition, and idea, and correlatively, to correct the meaning of nominalism and of realism.

It is not enough, indeed, to say that technical operation provides the paradigm of an essentially inductive thinking, whereas religious contemplation provides the model for a deductive theoretical thinking; this double paradigmatism is not limited to the sciences; it extends to philosophical reflection by providing it with the modes of knowledge that can be used and transposed onto other domains. Furthermore, the technical operation and religious contemplation provide the implicit axiomatic for all subsequent knowledge; there is indeed a link uniting the mode of knowledge (by concept, intuition, or idea) to the implicit axiomatic; this implicit axiomatic is constituted by the relation that exists between the reality to be known and the knowing subject, i.e., by the primary status of the reality to be known. Technical thought indeed provides the model for the intelligibility of the elements taken one by one, in their combination, and of their mutual relations that are constitutive of the ensemble; the real to be known resides at the end of the effort for knowledge, it is not a mass given all at once in its totality; made of elements because it is knowable as a combination of elements, this reality is essentially an object. Conversely, being the paradigm of deductive thought, religious thought starts from a function of the whole instantly recognized as having unconditional value, and which can only be made explicit, but not constructed and produced by the thinking subject. Religious thought provides the model for the contemplation of being, for a respect of being that can never fully resolve itself in knowledge, but for which a certain representation can be formed; the knowledge and the subject who receives it remain incomplete, inferior with respect to being. It is in fact being that is the true subject and the only complete subject. The subject of knowledge
319 is only a secondary subject, by reference to the first and by participation in it. Knowledge is conceived as an imperfect doubling of being, because the subject of knowledge is not the true subject. This contemplative mode of knowledge is the basis for idealist realism in philosophy; the εἶδος [eidos] is a view of being, a structure of being that exists for itself before being thought; it is not essentially, and from the start, an instrument of knowledge; it is first of all a structure of being; it is only secondarily and by way of participation that it becomes a representation in the soul, by virtue of a relation of kinship between the soul and the ideas; knowledge

is neither formed nor constructed by the subject; there is no genesis of knowledge, only the discovery of the real by the mind [*esprit*]. Knowledge is an imitation of being because being is essentially subject within itself, prior to any awareness by the secondary and imperfect subject that is man; for an example of such a metaphysical axiomatic, one can reference the one governing the theory of knowledge in Plato. The Good is absolute and first subject; it is what structures the plurality of ideas, none of which can be entirely subject for itself, insofar as it is this or that idea; the Good is the metaphysical translation of the function of totality as subject, prior to and higher than definite knowledge, guarantee of the intelligibility of this knowledge and of its validity; all knowledge is in a certain sense knowledge of the Good, not in itself directly, but indirectly and by reflection [*reflet*], for what makes knowledge be through the idea, is being's totality as one, absolute subject, toward which all effort of particular knowledge is a movement of ascendance. The knowledge of man carries out in the opposite direction the ontological path going from the Good to objects through ideas, going up from objects to the ideas of which they are the objects, and from the ideas to the Good, according to the analogical relation.

Operational knowledge, on the contrary, gives itself the possibility of construct-   320
ing its object; it dominates it and makes it appear, governing the genesis of its representation on the basis of manipulable elements, in the same way the artisan constructs the object he places before himself in order to assemble the pieces in a coherent way. The concept, as the instrument of operational knowledge, is itself the result of an operation of assembly, implying the process of abstraction and generalization, on the basis of an experience given in the particularity of the *hic et nunc*; the source of knowledge is here in the *hic et nunc*, rather than residing within unconditional totality and prior to all human gesture, governing even human gestures that are already conditioned by it before coming into existence and being fulfilled [*accompli*]. For contemplative knowledge, the real is absolute subject, whereas for operational knowledge it is still object, in the first sense of "that which is placed in front," as a piece of wood is placed on a workbench, waiting for its incorporation into the ensemble through the process of construction. For operational knowledge, the real does not precede the operation of knowledge; it comes after it. Even if it appears to precede it according to common experience, it succeeds it according to real knowledge, since this knowledge only grasps the real when it has reconstructed it through the manipulation of elements.

Now, this opposition between the two modes of knowledge is important, for the succession of philosophical schools shows that there are two currents of thought that cannot be allied, and which one can generally designate by the words of aposteriorism and apriorism; aposteriorism, which is empiricist, conceptualist, and partially nominalist (since knowledge, as it gains in abstraction, distances itself from elementary sources) defines knowledge as the operation that uses the concept; apriorism, on the contrary, being deductive, idealist, and realist unless it is a-cosmic, defines knowledge through the grasping of the real by means of the idea.

321   But, if the source of this opposition and incompatibility between the two basic metaphysical axiomatics were the split of the primitive mode of being in the world, into technics and religion, then one would have to affirm that philosophical knowledge cannot be content in grasping being by the concept or idea, not even successively by one or the other mode of knowledge. Philosophical knowledge, as the function of convergence, must call upon a mediate and higher mode of knowledge, reuniting concepts and ideas in its unity. Now, it is not entirely correct to identify intuition with the idea; knowledge by way of intuition is a grasping of being that is neither *a priori* nor *a posteriori*, but contemporaneous with the existence of the being it grasps, and which is at the same level as this being; it is not a knowledge by way of the idea, for intuition is not already contained within the structure of the known being; it does not belong to that being; it is not a concept, since it has an internal unity that grants its autonomy and its singularity, preventing a genesis through accumulation; lastly, knowledge by way of intuition is really mediate in the sense that it does not grasp being in its absolute totality, like the idea, or on the basis of elements and by combination, like the concept, but rather grasps being at the level of domains constituting a structured ensemble. Intuition is neither sensible nor intellectual; it is the analogy between the coming-into-being of known being and the coming-into-being of the subject, the coincidence of two comings-into-being: intuition is not merely the grasping of figural realities, like the concept, nor a reference to the totality of the ground of the real taken in its unity, like the idea; it aims at the real insofar as it forms systems in which a genesis occurs; it is the knowledge proper to genetic processes. Bergson made intuition the mode proper to the knowledge of coming-into-being; but one can generalize Bergson's method, without excluding a domain like matter from intuition, because it does

322   not appear to present the dynamic aspects necessary for intuitive comprehension; in fact, intuition can apply to every domain where genesis occurs, because it follows the genesis of beings, taking each being at its level of unity, without decomposing it into elements like conceptual knowledge, but also without destroying its identity by relativizing it with respect to the ground of a broader totality. The concept essentially retains from its technical nature the capacity to grasp figural realities; conversely, the idea is particularly apt for the knowledge of ground realities.

Intuition intervenes as mediator, considering ensembles in which there is genesis of structure, i.e., genesis of a correlation between figure and ground. Intuition is thus a particular procedure of philosophical knowledge, because by virtue of it thought can grasp being in its essence, which is the formula of its genetic coming-into-being, and remain at the neutral point of this coming-into-being in order to ensure the function of convergence.

For intuition, the level of unity is not totality, as with knowledge by way of idea, nor the element, as in conceptual knowledge. In this way, philosophical thought recovers a relation to being which was that of primitive magic, which then became that of aesthetic activity; known being, the world, is originally neither object nor subject; it is supposed an object when it is submitted to operational thought, as in scientific mechanistic thought; it is supposed a subject when it inspires contemplative knowledge, like the Cosmos of the Stoics; but the notion of object retains its technical origin, in the same way that the notion of subject retains its religious origin. Neither of the two fully apply to the world or to the human being, for they would only constitute a complete reality if they were taken together; in fact, the notion of object and the notion of subject are, by the very virtue of their origin, limits that philosophical thought must overcome, making knowledge according to the object and knowledge according to the subject converge within mediate knowledge, at the neutral point, according to intuition. Philosophical thought can thus 323 constitute itself only after having exhausted the possibilities of conceptual knowledge and knowledge by idea, which is to say after gaining a technical awareness and a religious awareness of the real; philosophy comes after technical construction and religious experience, and it defines itself as a capacity of intuition in the gap that separates them. Technics and religion are thus the two guiding poles that bring out the philosophical intuition of the real.

In philosophical thought, the relation between technics and religion is not dialectical; since, precisely to the extent that technics and religion are two opposite and complementary aspects of a primitive mode of being in the world, these two poles must be maintained together in the couple they form; they are simultaneous. An elucidation of philosophical problems can be valid only by accepting the unimodal character of thought coming from a single phase. The aesthetic vision of reality cannot satisfy philosophical research, for it only applies to chosen domains of the real, those in which the coincidence of figural realities and ground realities is possible without further elaboration. Aesthetic thought is not directly active; it does not reverberate on the real from which it starts; it limits itself to exploiting the real by detaching itself from it; it refracts aspects of reality, but it does not reflect them.

Conversely, philosophical thought goes further than aesthetic activity, since in taking genetic coming-into-being as its point of departure, philosophical thought reintegrates itself into genetic coming-into-being in order to fulfill it. Intuition is effectively a relation with the real that is both theoretical and practical; it knows and acts on it, because it grasps it in the moment in which it comes-into being; philosophical thought is thus philosophical gesture inserting itself into the reticular figure-ground structure that determines itself in being; philosophy intervenes as a power of structuration, as a capacity for the invention of the structures that resolve problems of coming-into-being, at the level of this intermediary nature between 324   the plurality and totality that is the reticular diversity of the domains of existence.

Intuition recovers the figural aspect and the ground aspect by in a real unity; for the elements and totality are not the concrete whole [ensemble] of being; the unity of being is the active center on the basis of which exist, through a split, figure and ground, i.e., the elements on the one hand and totality on the other; intuition knows and fulfills this unity of being, which is the combination of elements and totality; intuition is the relation of figure and ground in itself; it is not, like the idea, co-natural with the being it grasps, for this co-naturalness can grasp only the ground, which is not the whole [ensemble] of being, it is not abstract like the concept, which abandons the concreteness of being in order to preserve only the definite figure. Grasping the primitive relation of figure and ground, intuition is analogical with respect to being; it is a knowledge that justifies neither full realism nor pure nominalism, but a stable mixture of these two ways of envisioning knowledge's reach; intuition is not equivalent to being, it is not being as real idea, but it is analogical with respect to being, for intuition constitutes itself like being, through the same coming-into-being, which is relation of figure and ground. It recovers in being the complete existence of which magical thought was the presentiment, before the emergence of technics and religion. One can thus say that there are three types of intuition, according to the coming-into-being of thought; magical intuition, aesthetic intuition, and philosophical intuition. Aesthetic intuition is contemporary with the split of magical thought into technics and religion, it does not perform a veritable synthesis between the two opposite phases of thinking; it only indicates the necessity of a relation, and accomplishes it allusively in a limited domain. Conversely, philosophical thought must really accomplish this synthesis, 325   and it must construct culture as coextensive with the final result of all technical and religious thought; aesthetic thought is thus the model of culture, but it is not the entirety of culture; it is rather the announcement of culture, a demand for culture, rather than culture itself; for culture must combine in a real way the entirety of technical thought with the entirety of religious thought, and for this to happen it must be made up of philosophical intuitions, whose origin lies with the coupling carried out between concepts and ideas; aesthetic activity fills in the gap between technics and religion, whereas philosophical thought grasps and translates the extent of this interval; it considers it as positively significant and not as a statically free domain, but rather as a direction defined by the divergence of two modes of thought; whereas aesthetic thought is conditioned by coming-into-being, philosophical thought is born throughout divergent coming-into-being in order to make it re-converge.

The technicity of technical objects can thus exist at two different levels: original and primitive technical objects, which appeared as soon as magical thought ceased having an important functional signification, are indeed the real depositaries of technicity, as tools and instruments; but they are objects only to the extent that they can be put into action by a user; the user's gestures also belong to technical reality, even if they are contained in a living being that places its perceptive power, its functions of elaboration and invention, at the service of the technical task; the real unity is that of the task rather than that of the tool, but the task cannot be objectivated and can only be lived, experienced, accomplished, and not strictly speaking, reflected upon [*réfléchie*]. At the second level, technical objects are part of technical ensembles. Consequently, at the first level or at the second level, technical objects cannot be considered as absolute realities and as existing by 326 themselves, even after having been constructed. Their technicity can be understood only through the integration of the activity of a human user or the functioning of a technical ensemble. It would thus not be legitimate to seek to understand the technicity of an object on the basis of an induction comparable to that which one can apply to natural beings; the technical object, which never harbors all of technicity on its own, either because it is a tool or because it is the element in an ensemble, must be known by philosophical thought, i.e., by a thought that has the intuition of the coming-into-being of the modes of relation between man and the world.

The use of this genetic method defines the technical object through reference to the technicity of the artisanal operation or the technical ensemble, and not the technicity of the operation or that of the ensemble on the basis of some property of the object that technicity would be. This functional aspect, however, and this conditioning of the technical object's genesis are indeed effectively translated by a particular type of the technical object's coming-into-being, what we have called the concretization of the technical object. The process of this concretization can be directly apprehended by the examination of a certain number of examples of technical objects. But the sense of this concretization, which is an inherence in the object of a technicity that is not entirely contained in it, can be understood only by philosophical thought following the genesis of the technical and non-technical modes of the relation between of man and the world. Whence the use in this study of a genetic method applied first to technical objects and then to the study of the situation and role of technical thought in the whole [*l'ensemble*] of thought.[9]

---

9.   The last sentence of this paragraph, though added by Simondon to the original 1958 galley proofs, was left out of the 1958 edition, but included in the 2012 edition. — Ed.

CONCLUSION

To this day, the reality of the technical object has been relegated to the background behind the reality of human work. The technical object has been apprehended through human work, thought and judged as instrument, adjuvant, or product of work. However, one ought to be capable, in favor of man himself, to carry out a reversal that would enable what is human in the technical object to appear directly, without passing through the relation of work. It is work that must be known as a phase of technicity, not technicity as a phase of work, for it is technicity that is the whole of which work forms a part, and not the reverse.

A naturalistic definition of work is insufficient; to say that work is the exploitation of nature by men in society is to reduce work to an elaborate reaction by man, taken as a species, in confrontation with nature to which he adapts and which conditions him. What is at stake here is not a question of knowing whether this determinism in regard to the nature-man relation is one-way or contains reciprocity; the hypothesis of reciprocity does not change the basic schema, which is to say the schema of conditioning and the reactional aspect of work. In which case it is work that gives the technical object its meaning and not the technical object that gives its own meaning to work.

From the perspective offered here, work can be taken as an aspect of the technical operation, which is irreducible to work. There is work only when man must offer his organism as tool bearer, that is, when man must, along with his organism and his psychosomatic unity, follow the step-by-step unfolding of the human-nature relation. Work is the activity through which man actualizes the mediation between the human species and nature within himself; in this case we say that man operates as tool bearer because he acts on nature in this activity and follows this action, step by step, gesture by gesture. There is work when man cannot entrust the technical object with the function of mediation between the species and nature, and must fulfill this function of relation himself, through his body, his thought, his action.

Man thus lends his own individuality as a living being in order to organize this operation; it is in doing this that he is a tool bearer. On the other hand, when the technical object is concretized, the mixture of nature and man is constituted at the level of this object; operation on the technical being is not exactly work. Indeed, in work, man coincides with a reality that is not human, submits to this reality, and to a certain extent, slides between natural reality and human intention; in work, man models matter according to a form; with this form, which is the intention of a result, comes a predetermination of what must be obtained at the end of the work [*ouvrage*] in accordance with the pre-existing needs. This form-intention is not part of the matter onto which work applies itself; it expresses a utility or necessity for man, but it does not come from nature. The activity of work is what forms the link between natural matter and form, which comes from man; work is an activity that succeeds in making two realities as heterogeneous as matter and form coincide and renders them synergetic. And the activity of work makes man aware of the two terms he synthetically relates, because the worker must have his eyes fixed on these two terms which he must bring closer together (this is the norm of work), not on the interiority itself of the complex operation through which this bringing together is obtained. Work masks the relation in favor of the terms.

The servile condition of the worker has, moreover, often contributed to making the operation by which matter and form are made to coincide more obscure; the man who orders [*commande*] work to be done is concerned with what must appear in the given order [*ordre*], in terms of content and of the raw material that is the condition of execution, rather than with the operation that enables the process of taking form to occur: the attention is given to form and matter, not to the process of taking form as operation. The hylomorphic schema is thus a couple in which the two terms are clear and the relation obscure. Under this particular aspect the hylomorphic schema represents the transposition into philosophical thought of the technical operation reduced to work, and taken as a universal paradigm of the genesis of beings. It is indeed a technical experience, but a very incomplete technical experience that is at the basis of this paradigm. The generalized use of the hylomorphic schema in philosophy introduces an obscurity that comes from the insufficiency of this schema's technical basis.

Indeed, it is not enough to enter the workshop with the worker or slave, or even to take the mold into one's own hands and to operate the potter's wheel. The point of view of the working man is still too external to the process of taking form, which is the only thing that is technical in itself. It would be necessary to be able to enter the mold with the clay, to be both mold and clay at once, to live and feel their common operation in order to be able to think the process of taking form in itself. For the worker elaborates two technical half-chains that prepare the technical operation: he prepares the clay, makes it malleable, without lumps, without air 330 bubbles, and correlatively prepares the mold; he materializes the form by making it into a wooden mold, and makes matter pliable, capable of receiving information; then, he puts the clay into the mold and presses it; but it is the system constituted by the mold and the pressed clay that is the condition of the process of taking form; it is the clay that takes form according to the mold, not the worker who gives it its form. The working man prepares the mediation, but he doesn't fulfill [*accomplit*] it; it is the mediation that fulfills itself on its own once the conditions have been created; even though man is very close to this operation, he does not know it; his body pushes the mediation to fulfill itself, enables it to fulfill itself, but the representation of the technical operation does not appear in work. It is the essential part that is missing, the active center of the technical operation that remains veiled. For as long as man practiced work without using technical objects, technical knowledge could only be transmitted in an implicit and practical form, through professional habits and gestures: this motivating [*moteur*] knowledge is effectively what enables the elaboration of two technical half chains, the one starting from form and the one starting from matter. But it does not and cannot go further, it stops before the operation itself: it does not penetrate inside the mold. In its essence, it is pre-technical and not technical.

Technical knowledge, on the contrary, consists in starting from what happens inside the mold in order to find the different elaborations that can prepare it by starting from this center. Man cannot leave the center of operation in the dark, when he no longer intervenes as tool bearer; it is the center that must effectively be produced by the technical object, which does not think or feel, and which does not acquire habits. In order to construct the technical object that will function, man needs to represent to himself the way of functioning that coincides with technical operation, which accomplishes it. The functioning of the technical object belongs to the same order of reality, the same system of causes and effects, as the technical 331 operation; there is no longer heterogeneity between the preparation of the technical operation and the functioning of this operation; this operation prolongs the technical functioning just as the functioning anticipates this operation: the functioning is an operation and the operation a functioning. One cannot speak of the work of a machine, but only of its functioning, which is an ordered ensemble of operations. Form and matter, if they still exist, are at the same level and belong to the same system; there is continuity between the technical and the natural.

Making the technical object is no longer accompanied by this obscure zone between form and matter. Pre-technical knowledge is also pre-logical, in the sense that it constitutes a couple of terms without discovering the interiority of the relation (like in the hylomorphic schema). Technical knowledge on the contrary is logical, in the sense that it seeks the interiority of the relation.

It would be extremely important to observe that the paradigmatism arising from the relation of work, is very different from the one coming from the technical operation, from technical knowledge. The hylomorphic schema belongs to the content of our culture; it has been transmitted since classical Antiquity, and we often think of this schema as perfectly grounded, not relative to a particular experience, perhaps improperly generalized, but coextensive with universal reality. The process of taking form ought to be treated as a particular technical operation, rather than treating all the technical operations as particular cases of the process of taking form, which is itself obscurely known through work.

In this sense, the study of the mode of existence of technical objects should be extended by the study of the results of their functioning, and of man's attitudes in the face of technical objects. A phenomenology of the technical object would thus be extended into a psychology of the relation between man and the technical object. Yet, two pitfalls should be avoided in this study, and it is precisely the essence of the technical operation that makes them avoidable: technical activity belongs neither to the pure social domain nor the pure psychic domain. Technical activity is the model of the collective relationship, which cannot be confused with one of the two preceding ones; it is not the only mode and the only content of the collective, but it is of the collective, and, in certain cases, it is around technical activity that the collective group can arise.

What we mean here by social group is one that constitutes itself, like for animals, according to an adaptation to the conditions of the milieu; work is that through which the human being is mediator between nature and humanity as a species. On the opposite end, but at the same level, the inter-psychological relation puts individual before individual, establishing a reciprocity without mediation. Through technical activity, on the contrary, man creates mediations, and these mediations are detachable from the individual who produces and thinks them; the individual expresses himself in them, but does not adhere to them; the machine has a sort of impersonality which allows it to become an instrument for another man; the human reality that it crystallizes within itself is alienable, precisely because it is detachable. Work adheres to the worker, and reciprocally, through the intermediary of work, the worker adheres to the nature on which he operates. The technical object, which is thought and constructed by man, is not limited to simply creating a mediation between man and nature; it is a stable mixture of the human and the natural, it contains human and natural aspects; it gives its human content a structure comparable to that of natural objects, and allows for the integration of this human reality into the world of natural causes and effects. The relation of man to nature, rather than being only lived and practiced obscurely, takes on a status of stability, of consistency, making it a reality that has laws and an ordered perma- 333 nence. In edifying the world of technical objects and by generalizing the objective mediation between man and nature, technical activity re-attaches man to nature through a far richer and better defined link than that of the specific reaction of collective work. A convertibility of the human into the natural and of the natural into the human establishes itself through the technical schematism.

By thus constructing a structured world, the technical operation, rather than being pure empiricism, leads to the emergence of a new relative situation of man and nature. Perception corresponds to the direct challenge that the natural world puts to living man. Science corresponds to the same challenge through the technical universe. For work without obstacles, sensation is enough; perception corresponds to the problem that emerges at the level of work. On the other hand, as long as technics succeed, on the contrary, scientific thought is not called upon to emerge. When technics fail, science is near. Science corresponds to a problematic formulated at the level of technics, but unable to find a solution at the technical level. Technics intervenes between perception and science, in order to provoke a change of level; it provides the schemas, the representations, and the means of control of the mediations between man and nature. Having become detachable, the technical object can be grouped with other technical objects according to such or such setup [*montage*]: the technical world offers an indefinite availability of groupings and connections. For what takes place is a liberation of the human reality that is crystallized in the technical object; to construct a technical object is to prepare an availability. The industrial grouping is not the only one that can be brought about [*réalisé*] with technical objects: non-productive groupings can also be brought about, whose end is to relate man and nature through an ordered succession [*enchaînement réglé*] of organized mediations, to create a coupling between human thought and nature. 334 Here the technical world intervenes as a system of convertibility.

The work paradigm is what pushes us to the consideration of the technical object as a utilitarian one; the technical object does not carry its utilitarian aspect within itself as an essential definition; it is that which performs a determinate operation, which fulfills a certain function according to a determinate schema; but, precisely because of its detachable aspect, the technical object can be employed in an absolute manner as a link in a chain of causes and effects, without this object being affected by what happens at either end; the technical object can perform the analog of a work task, but it can also transport information beyond any utility for determinate production. It is the function, and not the work, that characterizes the technical object: thus, there are not two categories of technical objects, those serving utilitarian tasks and those serving knowledge; any technical object can be scientific and vice-versa; to the contrary, one could call scientific a simplified object that would only be suitable for teaching: it would be less perfect than the technical object. The hierarchical distinction of the manual and the intellectual does not affect the world of technical objects.

The technical object thus carries with it a broader category than that of work: operational functioning. This operational functioning presupposes, firstly, as a condition of possibility, an act of invention. Now, invention is not work; it does not presuppose the mediation between nature and the human species to be played out by somato-psychic man. Invention is not only an adaptive and defensive reaction; it is a mental operation, a mental functioning that is of the same order as scientific knowledge. There is an equality of levels between science and technical invention; the mental schema is what enables invention and science; it is the mental schema, once more, that allows the use of the technical object as productive, in an industrial ensemble, or as scientific, in an experimental setup. Technical thought is present in all technical activity, and technical thought is of the order of invention; it can be communicated; it authorizes participation.

Henceforth, above the social community of work and beyond the inter-individual relationship not supported by an operational activity, a mental and practical universe of technicity establishes itself, in which human beings communicate through what they invent. The technical object taken according to its essence, which is to say the technical object insofar as it has been invented, thought and willed, and taken up [*assumé*] by a human subject, becomes the medium [*le support*] and symbol of this relationship, which we would like to name *transindividual*. The technical object can be read as carrier of a definite information; if it is only used, employed,[10] and consequently enslaved, then it cannot bring any information, any more than a book that would be used as a wedge or pedestal. The technical object that is appreciated and known according to its essence, i.e., according to the human act that has founded it, penetrated it with functional intelligibility, valorized it according to its internal norms, carries with it pure information. One can call pure

---

10. In French, the primary meaning of *employer*, before its secondary one of "to employ" in the sense of to hire for work, is "to use." [TN]

information an information that is not evental, one that can be understood only if the subject receiving it solicits within itself a form analogous to the forms carried by the medium [*le support*] of information; what is known in the technical object is the form, the material crystallization of an operational schema and of a thought that has resolved a problem. In order for this form to be understood it is necessary that there be analogous forms in the subject: information is not an absolute advent, but the signification resulting from a relation of forms, one extrinsic and 336 the other intrinsic with respect to the subject. Hence, in order for the object to be received as technical and not only as useful, in order for it to be judged as the result of invention, as a carrier of information, and not as utensil, the subject receiving it must have technical forms within himself. An inter-human relation that is the model of *transindividuality* is thus created through the intermediary of the technical object. This can be understood as a relationship that does not relate individuals by means of their constituted individuality separating them from one another, nor by means of what is identical in every human subject, for instance the *a priori* forms of sensibility, but by means of this weight [*charge*] of pre-individual reality, this weight of nature that is preserved with the individual being, and which contains potentials and virtualities. The object that emerges from technical invention carries with it something of the being that has produced it, and from this being expresses what is least attached to the *hic et nunc*; one could say that there is something of human nature in the technical being, in the sense that this word "nature" could be used to designate the remainder of what is original, prior even to the humanity constituted in man; man invents by putting to work his own natural material [*support*], this ἄπειρον [ápeiron] which remains attached to each individual being. No anthropology taking as its starting point man as individual being can account for the transindividual technical relationship. Work, conceived as productive, insofar as it comes from the localized individual *hic et nunc*, cannot account for the invented technical being; it is not the individual who invents, it is the subject, vaster than the individual, richer than it, and having, in addition to the individuality of the individuated being, a certain weight of nature, of non-individuated being. The social group of functional solidarity, like the community of work, puts only individuated beings into relation. For this reason, it necessarily localizes and alienates them, even beyond all economic modality such as the one Marx describes under the name of capitalism: one could define a pre-capitalist alienation essential to work as work. Moreover, and symmetrically, the inter-individual psychological relation cannot put 337 anything other than constituted individuals into relation; rather than putting them into relation by means of their somatic functioning, as work does, it puts them into

relation at the level of certain ways of conscious, affective, representative functioning, and alienates them just as much. The alienation of work cannot be compensated
by way of another alienation, which would be that of the psychical detachment [*psy-
chique détaché*]: which is what explains the weakness of the psychological methods
applied to the problem of work and which want to resolve the problems by means
of mental functions. The problems of work are the problems having to do with the
alienation caused by work, and this alienation is not only economic, through the
play of surplus value; neither Marxism, nor this counter-Marxism, that is the psychologism in the study of work through human relations, can find the true solution,
because they place both sources of alienation outside of work, whereas work itself
insofar as it is work is the source of alienation. We don't mean to say that economic
alienation doesn't exist; but it is possible that the primary cause of alienation resides
essentially within work, and that the alienation described by Marx is only one of
the modalities of this alienation: the notion of alienation is worth generalizing in
order that one might situate the economic aspect of alienation; according to this
doctrine, economic alienation would already exist at the level of the superstructures,
and would presuppose a more implicit foundation, which is the alienation that is
essential to the situation of the individual at work.

    If this hypothesis is right, then the true path toward the reduction of alienation
would not be situated within the domain of the social (with the community of
work and class), nor in the domain of inter-individual relationships that social
psychology habitually envisages, but at the level of the transindividual collective.
338 The technical object made its appearance in a world in which social structures
and psychic contents were formed by work: the technical object thus entered into
the world of work, instead of creating a technical world with new structures. The
machine is thus known and used through work and not through technical knowledge; the relation of the worker to the machine is inadequate, because the worker
operates on the machine without his gesture continuing the activity of invention
in this gesture. The *obscure central zone* characteristic of work has transferred itself
to the utilization of the machine: it is now the functioning of the machine, the
provenance of the machine, the signification of what the machine does and the
way in which it is made that is the obscure zone. The primitive central obscurity of
the hylomorphic schema is preserved: man knows what goes into the machine and
what comes out, but not what happens in it: an operation takes place in the very
presence of the worker in which he does not participate, not even if he commands
or serves it. To command is still to remain external to what one commands, when
commanding consists in the activation according to a pre-established setup, made

for this activation, planned in order to operate this activation within the schema of the technical object's construction. The worker's alienation is translated by the break between technical knowledge and the exercise of the conditions of use. This break is so noticeable that the function of adjusting the machine is strictly distinct from that of the machine's user in a large number of factories, in other words, distinct from the worker, and it is prohibited for workers to adjust [*régler*] their own machines by themselves. The activity of adjustment, however, is the one that most naturally continues the function of invention and construction: adjustment is a perpetual, if limited, invention. The machine is not, indeed, thrown into existence once and for all from the moment of its construction, without the necessity of touch-ups, repairs, or adjustments. The original technical schema of invention is more or less properly realized within each produced unit [*exemplaire*], which is why 339 each unit more or less functions properly. It is by reference not to the materiality and particularity of each unit of a technical object, but by reference to the technical schema of invention that adjustments and reparations are possible and effective; what man receives is not the direct product of technical thought, but a unit of fabrication carried out on the basis of technical thought with more or less precision and perfection; this unit of fabrication is the symbol of technical thought, a carrier of forms that must encounter a subject in order to carry on and complete this accomplishment of technical thought. The user must have forms within himself so that, from the encounter of these technical forms with the forms carried by the machine, and more or less perfectly realized in it, a signification emerges on the basis of which the work done on a technical object becomes a technical activity and not simply work. The technical activity distinguishes itself from mere work, and from alienating work, in that technical activity comprises not only the use of the machine, but also a certain coefficient of attention to the technical functioning, maintenance, adjustment, and improvement of the machine, which continues the activity of invention and construction. The fundamental alienation resides in the break occuring between the ontogenesis of the technical object and the existence of this technical object. The genesis of the technical object must effectively be a part of its existence, and the relation of man to the technical object must contain this attention to the continued genesis of the technical object.

The technical objects that produce the greatest alienation are those meant for ignorant users. Such objects progressively deteriorate: they are new for a short time, and quickly begin to devalue when losing this aspect of being new because they can only distance themselves from the conditions of initial perfection. The sealing of delicate organs is indicative of this divide between the manufacturer, who is

340   identified with the inventor, and the user who only acquires usage of the technical object through an economic process; the warranty concretizes the purely economic aspect of this relation between manufacturer and user; in no way does the user continue the act of the manufacturer; through the warranty, the user purchases the right to force the manufacturer to return to the activity of manufacturing once again should the need become apparent. On the contrary, technical objects, that are not subject to such a status of separation between construction and use do not deteriorate over time; they are conceived so that the different organs constituting them can be continually replaced and repaired over the course of use: maintenance is not separate from construction, it continues it, and in certain cases, completes it, for instance by means of breaking it in [*rodage*], which is the prolongation and completion of construction through rectification of the surface conditions during their functioning. When the breaking-in cannot be performed by the user because of the limitations it imposes, then it must be done by the manufacturer after the assembly of the technical object, as is the case with plane engines.

The alienation resulting from the artificial divide between construction and use is thus perceptible not only by the man who uses the machine, or works on it, and cannot push his relationship with it beyond work; it also has repercussions within the economic and cultural conditions of the use of the machine and into the economic value of the machine in the form of a devaluation of the technical object, which happens more quickly the more accentuated this rupture is.

Economic concepts are insufficient to account for the alienation that is characteristic of work. It is in and of themselves that the attitudes of work are inadequate 341   to technical thought and technical activity, because what is lacking therein are the forms and the explicit mode of knowledge that are close to the sciences, and that would enable knowledge of the technical object. In order to reduce alienation, one would have to bring the aspect of work, of effort, of concrete application implying the use of the body, and of the interaction of function back to the unity of technical activity. It is, however, correct to say that the economic conditions amplify and stabilize this alienation: the technical object does not belong to the men who use it in industrial life. Furthermore, the relationship of property is very abstract, and it wouldn't be enough for workers simply to be the owners of their machines in order for alienation to be abruptly reduced; to possess a machine is not to know it. Non-possession, however, increases the distance between the worker and the machine on which his work is accomplished; it makes the relation even more fragile, more external, more precarious. It would have to be possible to discover a social and economic mode whereby the user of the technical object would not only be the owner of this machine, but also the man who chooses it and who maintains it. The worker, however, is placed in the presence of the machine without having chosen it; being placed in the presence of a machine is part of the conditions of employment, it is integrated into the socio-economic aspect of production. In an inverse sense, the machine is most often fabricated as an absolute technical object, functioning in itself, but poorly adapted to the exchange of information between the machine and man. *Human engineering*[11] does not go far enough in seeking to discover the best arrangement of command organs and control signals; this is indeed extremely useful research, and it is the point of departure for the search for the true conditions of the coupling between man and machine. But this research runs the risk of not being very effective unless it goes as far as the very foundation of the communication between man and machine. In order for information to be exchanged, man must possess within himself a technical culture, which is to say an ensemble of forms that, upon encountering the forms contributed by the 342 machine, will be able to elicit meaning [*signification*]. The machine remains one of the obscure zones of our civilization, at all social levels. This alienation exists as much at the management level [*dans la maîtrise*] as it does at the level of workers. The true center of industrial life, that in relation to which everything must order itself in accordance with functional norms, is technical activity. To ask oneself who owns the machine, who has the right to use [*employer*] new machines and who has the right to refuse them, is to turn the problem upside-down; the categories of capital and labor are inessential with respect to technical activity. The foundation of the norms and of law in the industrial domain is neither labor nor property, but technicity. Inter-human communication must establish itself at the level of technics through technical activity, not via the values of work or economic criteria; social conditions and economic factors cannot be harmonized because they are parts of different ensembles; they can only find mediation within a predominantly technical organization. This level of technical organization where man encounters man not as the member of a class but as a being who expresses himself within the technical object which is homogeneous with respect to his activity, is the level of the collective, going beyond the inter-individual and the given social.

The relation with the technical object cannot become adequate individual by individual, except in very rare and isolated cases; it can establish itself only to the extent that it will succeed in making this inter-individual collective reality, which we name transindividual, exist, because it creates a coupling between the inventive and organizational capacities of several subjects. There is a relation of

---

11. English in original. [TN]

causality and reciprocal conditioning between the existence of clear, non alienated technical objects used according to a status which does not alienate, and the constitution of such a transindividual relationship. It might be desirable for industrial life and companies to have technical committees at the level of their employee councils; in order to be efficient and creative, an employee committee should be essentially technical. The organization of channels of information in a company must follow the lines of technical operation and not that of social hierarchy or of purely inter-individual relations, which are inessential with respect to technical operation. The company, being the ensemble of technical objects and men, must be organized on the basis of its essential function, that is its technical functioning; it is at the level of the technical operation that the whole [ensemble] of the organization can be thought, not as a confrontation of classes, i.e., as a pure social ensemble, or as a grouping of individuals each having their psyche, which brings the ensemble down to an inter-psychological schema, but as a unit [une unité] of technical functioning. The technical world is a world of the collective, which is adequately thought neither on the basis of the brute social [fact], nor on the basis of the psyche. To consider technical activity as inessential in its very structure, and to take as essential either the social communities or the inter-human relations arising from technical activity, means not analyzing the nature of this very center of group and of inter-individual relationships, which is technical activity. To keep the notion of work as the center of the social, and the antagonistic permanence of a psychologism of human relations at the level of management and of capital, shows that technical activity is not thought for itself: it is approached only through sociological or economic concepts, studied as an occasion of inter-psychological relationships, but not grasped at the level of its real essence: the obscure zone subsists between capital and work, between psychologism and sociologism; developing between the individual and the social is the transindividual, which is, currently, not recognized and is studied through the two extreme aspects of either the work of the laborer or the management of the company.

The criterion of productivity [rendement], as well as the will to characterize technical activity through productivity cannot lead to a resolution of the problem; productivity is very abstract with respect to technical activity and does not allow one to enter into this activity in order to see its essence; several very different technical schemas can lead to identical levels of productivity; a number does not express a schema; the study of productivities and the means to improve them allows the obscurity of the technical zone to persist as completely as the hylomorphic schema does; it can only contribute to confusing the theoretical problems, even though it plays a practical role in the current structures.

But, philosophical thought can play a role in this elucidation of technical reality as an intermediary between the social and the individual psyche in the order of deontological problems. One cannot account for technical activity by classifying it among the practical needs of man, which is to say by allowing it to appear as a category of work. Bergson attached the technical activity to *homo faber*, and showed its relation with intelligence. But in this idea of the manipulation of solids as the foundation of technicity there is a presupposition that prevents the discovery of genuine technicity. Bergson in fact starts from the axiological dualism of the closed and the open, of the static and the dynamic, of work and reverie; work attaches man to the manipulation of solids, and the necessities of action are at the source [*au principe*] of an abstracting conceptualization, of the primacy given to the static with respect to the dynamic, to space with respect to time. The activity of work is thus enclosed well within materiality and attached to the body. This is so true that science itself, whose use of technical schemas Bergson sensed, is considered as having a practical and pragmatic function. In this sense, Bergson would be fairly 345 close to the broad trend of scientific nominalism, mixed with a certain pragmatism, which one can sense in Poincaré, and then in Le Roy, inspired by both Bergson and Poincaré. However, one might wonder whether this pragmatic and nominalist attitude toward the sciences is not based on an inexact analysis of technicity. In order to be able to affirm that the sciences have their sights set on the real, that they want the thing, it is not necessary to show that they have no relation with technics; for it is work that is pragmatic, not technical activity; the gesture of work is directed by its immediate utility. But technical activity reaches the real only at the end of a long process of elaboration; it rests on laws, it is not improvised; for technical solutions to be efficient, they must reach the real according to the laws of the real itself; in this sense, technics are objective despite all the aspects of utility they may present. Pragmatism is not wrong only because it incorrectly reduces the sciences to technics, whereas scientific knowledge emerges when technics fail before the real or fail to harmonize among themselves. Pragmatism is also wrong because it believes that it reduces science to a purely improvised solution by reducing it to technical activity. At root, pragmatism conflates work and technical operation.

In this sense the analysis of the mode of existence of technical objects therefore has an epistemological import. A doctrine like Bergson's opposes work to leisure and gives leisure, in the form of reverie, a fundamental epistemological privilege: this opposition returns to the one made by the Ancients between servile occupations and the liberal, disinterested occupations, having the value of pure knowledge, whereas servile occupations only had a value of utility. Pragmatism, by appearing to reverse the hierarchy of values, defines the true by the useful; but it preserves the schema of opposition between the norm of utility and the norm of truth, to such an extent that it results in a relativism in the order of knowledge, or at nominalism if this attitude is pushed to its most rigorous and extreme consequences; science is not more true, but more useful for action than common perception.

If, on the contrary, one appeals to the veritable mediation between nature and man, namely to technics and to the world of technical objects, then one arrives at a theory of knowledge that is no longer nominalist. It is through operation that a becoming aware takes place, but *operational* is not synonymous with *practical*; the technical operation is not arbitrary, pliable in every way to the whims of the subject according to the randomness of immediate utility; the technical operation is a pure operation that puts into play the veritable laws of natural reality; the artificial is something natural that has been solicited, not something false or human that has been mistaken for something natural. In Antiquity, the opposition between operational knowledge and contemplative knowledge valorized contemplation, and the σχολή [skholē] that conditioned it. Technics, however, is neither work nor σχολή [skholē]. Philosophical thought, insofar as it comes from the tradition and uses schemas coming from the tradition, does not contain any reference to this intermediary reality between work and σχολή [skholē]. Axiological thought itself is at two levels and reflects this opposition between work and contemplation; the notions of the *theoretical* and the *practical* still refer to this adversative distinction. In this sense, it is permissible to think that the dualism inherent in philosophical thought, a dualism of principles and attitudes because of the double reference to the theoretical and the practical, will be profoundly modified by the introduction of technical activity considered as an area of reflection within philosophical thought. Bergson has only reversed the correspondences of σχολή [skholē] and work, by granting work the function of establishing a relationship with solids, hence with what is static, whereas the Ancients considered it a fall into the world of generation and corruption, hence of coming-into-being; Bergson, inversely, attributes to σχολή [skholē] the power to allow a coincidence with duration, with the moving, whereas the Ancients assigned to contemplation the role of enabling knowledge of the eternal. But this reversal does not change the condition of duality and the devaluation of the term corresponding to the work of man, whether this term is the moving or the static. It seems that this opposition between action and contemplation, between the immutable and the moving, must cease in the face of the introduction of the technical operation within philosophical thought as area of reflection and even as paradigm.

GLOSSARY OF TECHNICAL TERMS

Rocker switches. Component part with two states of equilibrium. When the two states of equilibrium are stable, the rocker is called indifferent; if it has both a stable state of equilibrium and an unstable state of equilibrium, it is called monostable: it goes from the stable state to the unstable or almost stable under the influence of an external signal; if the monostable rocker spontaneously returns to the stable state as soon as the signal disappears, then it is only called a monostable switch; if on the contrary, after the disappearance of the signal, the quasi-stable state continues for a time whose duration is determined by the characteristics of the circuit, then the setup is called a differentiated monostable rocker.

The Eccles-Jordan circuit constitutes an indifferent rocker; two identical triodes are coupled such that one is blocked (non-conductive as a result of a strong negative polarization of its command grid) while the other is conductive: a fraction of the anodic potential of each triode is transmitted to the other triode's grid through a resistance. The external signals arrive indistinctly on both anodes and are transmitted to the grids by the bridge dividing the resistances and by the condensers. These signals, in the form of negative pulses, do not act on the blocked triode, but modify the state of the conducting triode, which leads to the switching of the setup: the previously conductive triode becomes non-conductive, and the non-conductive triode becomes conductive. This circuit is often used in calculating machines, because it delivers only one of the two pulses it receives, this pulse in turn is capable of activating another ensemble of triodes; it therefore realizes through its physical functioning the analog of the mental operation of addition. To establish a chain of Eccles-Jordan circuits, a counting scale using a numeric system on the basis of 2 has to be built. In its pure form, the counting scale is used at the output of pulse counters, and more specifically in the measurement of radioactivity; when integrated into more complex setups, it provides the basis for binary electronic calculating machines. It is possible to construct mechanical rockers: however, the electronic rocker presents a considerable advantage, which is that of functioning speed (100,000 state changes per second).

Amplification class. Amplification classes are defined by the classes of functioning of the electronic tubes, which perform this function; the class corresponds to the position of the point of functioning on the characteristic of the anodic current in relation to the command grid's voltage; in class A, the point of functioning moves without leaving the rectilinear part of the characteristic; in class B, the grid receives a negative polarization such that the anodic current remains null in the absence of variable voltage on the grid; in class C, the grid receives an even stronger polarization. Under these conditions, in class A, an average signal does not noticeably change the average anodic throughput; but, if the signal increases, with a light bulb mounted in automatic polarization through the insertion of a resistor in the cathode, then the resulting increase in polarization diminishes the lamp's slope, which constitutes a negative reaction.

Electrolytic condenser. A condenser constituted by two electrodes immersed in an electrolytic liquid which, electrolyzed by the flow of a current, deposits a thin layer of insulation onto one of the electrodes; the liquid thus constitutes one of the plates, separated from the electrode covered by the insulating layer that plays the role of a dielectric. Left to its own devices, the condenser loses its dielectric, but the latter reforms after the current has flowed for a given period of time. This type of condenser enables the collecting of a rather large quantity of energy in a rather small volume, due to the thinness of the insulating layer; it has a maximum voltage of utilization (550 to 600 volts) and is characterized by more important losses than those of a condenser with a dry and permanent dielectric, such as mica or paper).

Converter. An ensemble constituted by an electric motor and a mechanically coupled generator. In contrast with the converter, the commutator uses a single rotor, which creates, in addition to the mechanical coupling, a magnetic coupling between the two coils, preventing the conversion of alternating current into continuous current, whereas the converter, despite a lower output, can perform this conversion.

Detonation and deflagration. Detonation is a combustion that, at the heart of an explosive mixture, occurs in an extremely brief time and at all points of the volume at the same instant. Deflagration on the other hand is a rapid but progressive combustion, starting in one point and propagating step by step across the entire volume, in the form of an explosive wave, like the burning of a trail of [explosive] powder that one ignites at one end.

Detonation is generally conditioned by a global state of the system (temperature, pressure) acting on all the gas molecules at the same instant, whereas deflagration must be triggered at a single point. Detonation exerts a destructive rupture effect; it is this that one seeks to obtain in dynamite through a percussion that creates a state of pressure on the entire mass of the charge at the same instant (the trigger has a mercury fulminate, whose purpose is not to ignite, but to compress); an explosive charge, lit at a single point, deflagrates rather than detonating. In an engine, combustion must be triggered before the global state of temperature and pressure provokes a detonation, causing the phenomenon called clatter.

Magneto. Complex electric machine, composed of one or several fixed magnets, which create a magnetic field where two coils are wound up together around a steel core rotate. The first coil is made of thick wire (like the primary coil of an induction coil), is short circuited through an external switch by the axis of the steel core; this switch opens the moment the variation of flux through the core is at its maximum, i.e., the moment where the current is most intense in the primary winding. The abrupt variation in intensity caused by this break in the primary winding creates a peak of elevated voltage in the secondary coil made of thick long wire, playing the role of the secondary coil of an induction coil. This high voltage pulse, distributed by a turning distributor on either spark igniter, creates a spark between the electrodes of the light bulb.

The magneto thus produces both low voltage energy and the high intensity of the primary, like a magnetic alternator, and high voltage pulse, like a Ruhmkorff coil (pulse transformer); lastly, it is the revolving of the axis that commands the rupture causing the variation of voltage in the primary circuit; it is this revolving that once more activates the distributor, alternately sending the peak of high voltage unto the spark igniters during the course of the igniting circuit. In addition to its concrete aspect, the magneto presents the following advantage: the higher the speed of rotation, the greater the variation in magnetic flux in the core will be, giving rise to a homeostatic effect: ignition is more energetic at high rather than low operating levels, which compensates for the greater difficulty of a correct ignition at higher operating levels following the mixing of the carburized mix in the cylinder; on the contrary, when lit by a battery and an induction coil, the energy available for the primary coil decreases as the engine speed increases, because of the phenomena of self-induction of the primary coil, opposing themselves to the sufficiently rapid establishing of curing in the primary. However, because of the plurifunctional role of its components, the magneto requires a high quality manufacturer.

Magnetostriction. Variations in the volume of a piece of metal under the influence of a magnetic field; iron and nickel present important properties of magnetostriction. If the magnetic field is alternating, a mechanical vibration results. This phenomenon is used to construct an electromechanical converter suited to high frequencies (ultrasound generators); it interferes with the transformers in oscillators, because the vibrations produced by the magnetic circuit's plates are transmitted into the framework and create a sound that is difficult to suppress.

Curie temperature. Temperature above which magnetization is unstable: ferromagnetic substances abruptly become paramagnetic; for iron, the Curie temperature is approximately 775° C.

Relaxant. Setup of natural ensemble that is the seat of a phenomenon of relaxation. The phenomenon of relaxation is a non-oscillating iterative functioning (repeating itself for an indefinite number of times and in a regular way); in relaxation, at the end of a cycle, i.e., the state of the system at the end of the cycle, it triggers the beginning of the cycle by starting a defined phenomenon: thus there is discontinuity from one cycle to the next; when a cycle is started, it continues on its own, but each cycle, in order to produce itself, requires the completion of the preceding one. Such is the functioning of intermittent fountains: the siphon begins, which entails the flowing of a certain quantity of liquid; then the siphon stops and will not recommence until the level of water has reached a certain height. A hydraulic ram functions by relaxation. On the contrary, in oscillation this critical phase of recommencement of the cycle is absent, but a continuous transformation of energy, for instance of potential energy into kinetic energy in the heavy pendulum, or of electrostatic energy into electro-dynamic energy, in an oscillating circuit with self-inductance and capacity. Oscillators operate in a sinusoidal manner, whereas relaxers do so in a "saw tooth" pattern. In fact, only oscillators have a real oscillation period; relaxants have a period only in relation to well defined magnitudes, for instance the amount of energy expended in each cycle; any variation in these magnitudes entails a variation of the duration of the cycle; whereas oscillators have a period defined by the characteristics of the setup itself. The confusion between an oscillator and a relaxant comes from the need to plan for the system's maintenance of the oscillations which call upon devices functioning as relaxants; hence, if one inserts a triode into a circuit with self-inductance and the capacity to maintain the oscillations, one can no longer obtain rigorously sinusoidal oscillations; one must then choose between obtaining a low level of almost sinusoidal oscillations and the production of a high level of oscillations distancing itself notably from the sinusoid, which require a strong coupling between the oscillating system and the maintenance system; at the same time as this coupling increases, the behavior moves toward that of a relaxant, with greater dependency of the frequency on external conditions (in particular the amount of energy dissipated during each cycle). The relaxation oscillator, which does not have an issue such as kinetic energy (inertia) is very easy to adjust; thus, a thyratron mounted in a system with resistances and capacitors can be adjusted by a variation of voltage on the control gate, determining the critical point of recommencement of the cycle. A genuine oscillator, on the other hand, is less easily adjustable and synchronizable: it is more autonomous, as one can see from pilot-oscillators, possessing a weak coupling of the maintenance circuit with the oscillating circuit, and a low level of output. Bodies that are both elastic and piezo-electric, like quartz, provide excellent oscillating circuits; a vibrating blade, a diapason can also provide oscillating systems that can be self-maintained.

Thermo-siphon. A device transporting heat, for heating or cooling, using the fact that water dilates, and consequently becomes lighter as it heats up; water becomes lighter and rises within the hot half of the circuit, while it becomes denser and descends and returns into the hot source. The circulation is all the more rapid as the difference in temperature between the hot source and cold source becomes larger: this system is thus homeostatic. However, because of the slowness of the circulation of water, it requires a more cumbersome and heavier apparatus than that used by a pump.

Synchronization pip. Brief signals enabling the enslaving of a device with recurrent functioning to a pilot device. When the pilot device is a sinusoidal oscillator, a brief signal, whose phase is well determined (for instance by capping the voltage of oscillation), is extracted prior to this oscillation. French television broadcasting standards place synchronization signals in the infra-black, below the voltage at which the electromagnetic beam of the cathodic ray tube becomes extinct, such that they can be transmitted on the same frequency as that which carries the modulation of the image without disturbing the latter: the passage from one line to the next, or from one image to the next, is translated only by a complete extinction of the dot on the screen.

BIBLIOGRAPHY

This bibliography does not include the titles of philosophical works that have become classics and integrated in the history of thought, nor the titles of the numerous technical studies used as sources in specialized journals, but only those of technological works and works of information theory, of cybernetics and the philosophy of technology, which are of a documentary nature.

Ashby, William Ross, Grey Walter, Mary A. B. Brazier, W. Russel Brain et al. Trans. J. Cabaret. *Perspectives cybernétiques en neurophysiologie.* Paris: Presses universitaires de France, 1951.

Biologie et cybernétique. Cahiers Laënnec 2. Paris: Lethielleux, 1954.

Canguilhem, Georges. *La connaissance de la Vie.* Paris: Hachette, 1952. *Knowledge of Life.* Translated by Stefano Geroulanos and Daniela Ginsburg. New York: Fordham University Press, 2008.

Chapanis, Alphonse. *Research Technics in Human Engineering.* Baltimore: Johns Hopkins University Press, 1959.

Chrétien, Lucien. *Les machines à calculer électroniques.* Paris: Chiron, 1951.

Colombani, P., G. Lehmann, J. Loeb, A. Pommelet and F. H. Raymond. *Analyse, synthèse, et position actuelle de la question des servomécanismes.* Paris: Société d'édition d'Enseignement supérieur, 1949.

Couffignal, Louis. *Les machines à calculer, leurs principes, leur évolution.* Paris: Gauthier-Villars, 1933.

Couffignal, Louis. *Les machines à penser.* Paris: Minuit, 1952.

Daumas, Maurice. *Les instruments scientifiques aux XVIIe et XVIIIe siècles.* Paris: Presses universitaires de France, 1953.

Dictionnaire de l'industrie, ou: Collection raisonnée des procédés utiles dans les Sciences et dans les Arts. Paris: Rémont, 1795.

Recueil de Planches sur les Sciences, les Arts libéraux et les Arts méchaniques, avec leur explication (for the *Encyclopédie*). Paris: Briasson, Le Breton and Durand, 1762.

Diels, Hermann. *Antike Technik.* Leipzig and Berlin: Reclam, 1924.

Diesel, Eugen. *Das Phänomen der Technik.* Leipzig: Reclam, 1939.

Encyclopédie moderne. Paris: Firmin Didot, 1846.

Foerster, Heinz and Claus Pias. *Cybernetics: The Macy Conferences 1946–1953: The complete transactions.* Berlin: Diaphanes, 2016.

Friedmann, Georges. *Le travail en miettes.* Paris: Gallimard, 1956.

Gellhorn, E. *Physiological Foundations of Neurology and Psychiatry.* Minneapolis: University of Minnesota Press, 1953.

Jackson, Willis, et al. *Communication Theory.* London: Butterworth, 1953.

L'Encyclopédie et le Progrès des Sciences et des Techniques. Paris: Centre International de Synthèse, Presses universitaires de France, 1951.

La Grande Encyclopédie. Paris: Lamirault et Cie., 1886–1902.

Latil, Pierre de. *La Pensée artificielle.* Paris: Gallimard, 1953; *Thinking by Machine: A Study of Cybernetics.* Trans. Y. M. Golla. Boston: Houghton-Mifflin, 1957.

Le Cœur, Charles. *Le rite et l'outil.* Paris: Presses universitaires de France, 1939.

Leroi-Gourhan, André. *L'Homme et la matière.* Paris: Albin-Michel, 1943.

———. *Milieu et Techniques.* Paris: Albin-Michel, 1945.

Loeb, Julien (editor), Louis de Broglie, et al. *La Cybernétique: Théorie du signal et de l'information.* Paris: Éditions de la Revue d'Optique Théorique et Instrumentale, 1951.

Ménard, René and Maurice Sauvageot. *Le travail dans l'Antiquité: Agriculture, industrie.* Paris: Flammarion, 1913.

Ombredane, André and Jean-Marie Faverge. *L'analyse du travail: Facteur d'économie humaine et de productivité.* Paris: Presses universitaires de France, 1955.

Privat-Deschanel. *Traité élémentaire de physique.* Paris: Hachette, 1869. *Elementary Treatise on Natural Philosophy.* Trans. J. D. Everett. London: Blackie & Son, 1978.

Recueil de Planches pour l'Encyclopédie. Paris: Panckoucke, 1793.

Schröter, Manfred. *Philosophie der Technik.* Munich: Oldenbourg, 1934.

Schuhl, Pierre-Maxime. *Machinisme et philosophie.* Paris: Presses universitaires de France, 1946-1948.

Slingo, William and Arthur Brooker. *Electrical Engineering.* New York: Longmans, Green & Co., 1900.

Strehl, Rolf. *Die Roboter sind unter uns.* Oldenburg: Gerhard Stalling, 1952. *The Robots Are Among Us.* Trans. Herman Scott. London/New York: Arco, 1955.

Structure et Évolution des Techniques 39–40 (July 1954–January 1955).

Tétry, Andrée. *Les Outils chez les êtres vivants.* Paris: Gallimard, 1948.

The Complete Book of Motor-Cars, Railways, Ships, and Aeroplanes. London: Odhams Press, 1949.

Tucker, D. G. *Modulators and Frequency Changers.* London: Macdonald, 1953.

Van Lier, Henri. *Le nouvel age.* Paris: Casterman, 1962.

Walter, Grey, W. *The Living Brain.* London: Duckworth, 1953.

Wiener, Norbert. *Cybernetics or Control and Communication in the Animal and the Machine.* New York: John Wiley & Sons, 1946.

———. *The Human Use of Human Beings: Cybernetics and Society.* Boston: Houghton & Mifflin, 1950.

Gilbert Simondon (1924–1989) was a French philosopher of technology best known for his theory of individuation through transduction in a metastable environment. Simondon's work has been championed by thinkers such as Gilles Deleuze and continues to attract new intersest within a variety of academic fields.